GUNS FOR THE
TSAR

GUNS

NORTHERN ILLINOIS UNIVERSITY PRESS

DEKALB, ILLINOIS   1990

FOR THE
TSAR

**American
Technology
and the
Small Arms
Industry in
Nineteenth-
Century
Russia**

Joseph Bradley

© 1990 by Northern Illinois University Press
Published by the Northern Illinois University Press, DeKalb, Illinois 60115
Design by Julia Fauci

Library of Congress Cataloging-in-Publication Data

Bradley, Joseph.
Guns for the Tsar : American technology and the
small arms industry in nineteenth-century Russia /
Joseph Bradley.
p. cm.
Includes bibliographical references.
ISBN 0–87580–154–4 (alk. paper)
1. Firearms industry and trade—Soviet Union—
History—19th century. 2. Technology transfer—Soviet
Union—History—19th century. I. Title.
HD9744.F553S683   1990
338.4'762344'094709034—dc20   90–34654
CIP

To my mother
and in memory of my grandparents
and of Bob

Contents

Acknowledgments

● FEW AUTHORS END UP WRITING the book they planned to write, and this book has become quite different from its original intent. Working as an editor in a Moscow publishing house during the years of detente in the mid-1970s, I became interested in the question of Russian-American business contact in earlier times. After my return to the United States, I heard that the Weatherhead Foundation was interested in supporting a study of Samuel Colt's attempts to sell revolvers to the Russian government. It seems that Al Weatherhead, trustee of the foundation and a gun collector, had purchased a Colt model revolver and wanted to know why it had the name of a Russian factory on the barrel. My interests matched those of the foundation, and I was awarded a grant. Once I began to investigate Colt's connection with the Russian army, I soon realized that Russian-made Colt revolvers were just a small part of a much larger transfer of American small arms technology. In order to assess its full import, it was necessary to recast the project and to study this technology transfer in the context of Russia's changing military needs and productive capabilities. Samuel Colt and other American weapons manufacturers not only sold Russia rifles and revolvers; they also supplied Russia with modern machinery, machine tools, research and design facilities, and production organization and supervision.

This book could never have been written without the generous help of a variety of individuals and institutions. The Weatherhead Foundation provided the financial support to get this project off the ground. Donald Price, former dean of Harvard's Kennedy School of Government, and Thane Gustafson brought this source of financial support to my attention. The Weatherhead grant was administered by the Russian

Research Center of Harvard University. I am grateful to Abram Bergson, the director when my research began, and to Ned Keenan, Adam Ulam, Marshall Goldman, and Mary Towle, for their encouragement in the early stages and their patience when several delays postponed completion of the book. I am also grateful to the International Research and Exchanges Board, the Oklahoma Foundation for the Humanities, and the Office of Research and the Faculty Summer Research program at the University of Tulsa for additional financial support.

Several institutions provided congenial settings for research or writing. In addition to the Russian Research Center and the University of Tulsa, three visiting appointments facilitated access to collections, and I am grateful to the School of Slavonic and East European Studies of the University of London and the history departments of Ohio State and Georgetown universities. I have also benefited from the pleasant meeting ground of the Summer Slavic Workshop at the University of Illinois.

Army and armory sources, as well as business records, are often inaccessible. When they are accessible, they are often reticent. Although unpublished records in Soviet archives were not available to me, the development of the small arms industry can be pieced together from published accounts. I have been issued many library cards. I am grateful to the Slavic Reference Library at the University of Illinois, and particularly to Helen Sullivan and Larry Miller for their tireless assistance in pursuing obscure sources. I would also like to thank J. S. G. Simmons of Oxford University and the staffs of the Library of Congress, the British Library, the New York Public Library, the Lenin library, the Leningrad Public Library, the Smithsonian Institution Libraries, the libraries of Harvard, Columbia, Helsinki, and Ohio State universities, and the Interlibrary Loan office at the University of Tulsa. Curators and archivists of a variety of institutions made their collections available to me or assisted me in pursuing unpublished or pictorial sources: the National Archives; the Department of Prints and Photographs of the Library of Congress; the Division of Armed Forces History of the National Museum of American History; the Connecticut State Library, the Connecticut Historical Society, and the Wadsworth Atheneum in Hartford; the Public Record Office, the National Maritime Museum, the Royal Artillery Institution, the National Army Museum, the Tower of London Armouries, and the Imperial War Museum in London; the Weapons Museum of the School of Infantry in Warminster; the American Precision Museum in Windsor, Vermont; the Hagley Museum and Library in Wilmington; the Yale University Library; and the West Yorkshire Archive Service in Leeds.

I have benefited greatly from many individuals who shared their technical expertise regarding firearms, machinery, technology transfer, and international business, in person or by correspondence. In the early stages of the project Larry Wilson, Joe Rosa, and Roy Jinks helped guide me through the intricacies of the weapons and of the foreign business of the Colt and Smith and Wesson companies. Edwin A. Battison has clarified several points regarding gunworking machinery. I am also grateful to Leonid Tarassuk, Runo Kurko, Fred Carstensen, Walther Kirchner, David Landes, Merritt Roe Smith, Herbert Houze, William E. Meuse, Herb Woodend, William B. Edwards, Harry Hunter, and Roy Marcot for sharing information and technical expertise or for directing me to important sources.

Over the years I have had much research assistance. Irina Krivtsova helped me find references in many obscure journals and provided encouragement for many years. I am also grateful to Paul Josephson, Susan Matula, Debbi Schwartz, Linda Kottum, Robert Gross, Linda Khan, and Alexander Stolarow. Kristi Treat helped type an earlier draft. Mary Lincoln and the editors of Northern Illinois Press have acted with grace and patience.

Walter Pintner, Bruce Lincoln, Jake Kipp, John Bushnell, and Hans Rogger read earlier drafts and offered many helpful suggestions. Larry Cress took time from his duties as department chair to offer suggestions from the perspective of an American historian. The anonymous reviewer at Northern Illinois University Press suggested helpful organizational changes. Finally, Chris Ruane not only offered incisive criticism and constructive suggestions, she also provided encouragement at a critical stage and convinced me that this study could speak to a broader audience. The shortcomings that remain are solely mine.

GUNS FOR THE TSAR

So they put the flea under the microscope as the left-handed craftsman had directed, and no sooner had the Emperor taken a look at it through the glass than a smile broke out on his face and he took the Tula craftsman just as he was, unkempt and covered with dust, and embraced and kissed him, and then he turned to all his courtiers and said:

"You see, I knew better than any of you that my Russians would never let me down. Why, the fellows have shod the flea's feet."

> —Nikolai Leskov, "The Left-handed Craftsman: A
> Tale of the Cross-eyed, Left-handed Craftsman of
> Tula and the Steel Flea"

In other countries, uniformity of lock parts has been deemed impossible even after many and lengthy attempts. . . . But once I saw at the Tula factory a large number of locks which had been disassembled in the inspection shop and whose parts had been mixed. When these parts were reassembled, they fit together so precisely, it was as if they had been deliberately fitted.

> —Iosif K. Gamel',
> Opisanie Tul'skogo oruzheinogo zavoda v istor-
> icheskom i tekhnicheskom otnoshenii

We borrowed the machine method of making firearms chiefly from America, where interchangeable wares are made with such precision that one by one our officers, seeing this production for the first time, not only brought back their full approval, but also the conviction that such precise work was still unthinkable here at home.

> —Miasoedov, "Sravnitel'noe opisanie ruzhei Ber-
> dana 2, Mauzera, Gra, i Gochkisa"

INTRODUCTION
The Arms Industry and Development

●   THIS CASE STUDY EXAMINES TECHNOLOGICAL change and modernization in one Russian industry—the nineteenth-century small arms industry. This process of modernization contains a paradox. As the storyteller Nikolai Leskov tells us, only partly tongue in cheek, Tula craftsmen in the age of Alexander I were superior to their English cousins: they could shoe the toy steel fleas made in England. But the Tula armory, the largest of three government small arms factories, boasted not only of its fine craftsmen. Several contemporary sources indicate that in the first half of the nineteenth century, Tula was well equipped with gunmaking machinery capable of producing weapons with interchangeable parts. English mechanics testifying before a parliamentary committee referred to the production of uniform gun parts as the "Russian plan." Prior to the Crimean War, Russia was virtually self-sufficient in the production of military arms.

Yet at a time of rapid changes in the design and manufacture of infantry rifles, Russia's domestic small arms industry had become technologically stagnant. Searching abroad for improvements in weapon designs, Russian officers were incredulous at the sight of modern American machinery and precision work. Influenced by American armory production, Russia borrowed the machine method of making firearms from the United States and mechanized its three government armories. The paradox of borrowing abroad a technology that had been available earlier at home raises two questions. First, why did Russia need to import a foreign technology to modernize its small arms industry in the 1870s? Second, and perhaps more important, why did a native technology fail to reproduce itself and keep up with foreign innovations? This book attempts to answer these questions.

Nineteenth-century Russia provides the first example of a developing society in the modern sense: a traditional social structure, a self-sufficient economy, and an entrenched state apparatus.[1] For centuries the needs of war had shaped the policies of the Russian service state. In the eighteenth and nineteenth centuries, the socioeconomic and military system bequeathed by Peter I and his Romanov predecessors mobilized the resources of a large and poor nation. The Russian empire expanded at the expense of Sweden, Poland, and Turkey, and its armies chased Napoleon to the heart of Europe. Russia helped suppress nationalist revolt in central Europe and became the bulwark of the old order in Europe. For all intents and purposes the autocracy was a veritable Leviathan and the army was unchallenged. Yet the army that had defeated Napoleon could not defend positions in Crimea forty years later.

"A nation in arms demands a nation of armorers and technicians to sustain and maintain it," wrote J. F. C. Fuller.[2] Despite the importance of the military and the exigencies of war and empire, Russia was not a nation in arms, much less a nation of armorers. The state's military needs, objectives, and means; the theory and tactics of warfare; the training and skills of the peasant soldier; the weapons systems available; and the craft methods of manufacturing—all composed a matrix that had served the old regime well. During the first half of the nineteenth century, a large standing army, with its small annual levies of serfs, had satisfied Russia's military and social needs. According to statute, a soldier achieved freedom from bondage on completion of military service. However, few serfs recruited for twenty-five years of service survived to become free men. In Europe, meanwhile, the era of universal military conscription had already opened: trained reserves could quickly turn a relatively small standing army into a "nation in arms." Were Russia to have inducted large numbers of serfs for short periods of active duty, millions of serfs would have earned their freedom, thereby destabilizing the nation's social and, it was feared, political order. Therefore, the creation of a modern mass army, long resisted by Alexander I and Nicholas I, depended on a profound social and legal change—the abolition of serfdom.[3]

Russia's defeat in the Crimean War strained the relationship between army and society. In its first issue in 1858, a military monthly editorialized, "Nowhere is there such passion in opinions, such an irritated tone in references to the past, as in Russia."[4] When Alexander II ascended the throne in 1855, it became apparent that the days of serfdom were numbered. The Great Reforms launched by Alexander II and by enlightened bureaucrats such as Minister of War Dmitrii Miliutin freed the serfs and introduced universal military service. New relationships between military and society began to replace the old. Laborer and soldier both were now free. For the latter, virtual lifetime duty was replaced by six years of active duty and nine years of reserve duty. More important, if less tangible, the Russian government relinquished its monopoly of control over many areas of life in an effort to revitalize the nation and the state.

A large standing army in the era of smoothbore muskets had experienced little change in tactical doctrine. Based on the scriptures of Suvorov, Russian tactical doctrine emphasized action in mass and bayonet force over open formations and firepower. With the introduction of the rifle, however, western tactical doctrine more and more recognized initiative, flexibility, mobility, agility, action in open formation, and adaption to surroundings. But after Crimea, according to the same Russian military journal cited above, "In all the articles published on this subject, and especially in conversation, there has been no mercy shown to our previous regulations or to our rules of tactics."[5] Although the army eventually surmounted many of the tactical and organizational challenges posed by the appearance of long-range breechloading rifles, as Bruce Menning argues, tactical theory and practice continued to stress the importance of morale and the bayonet charge.[6]

Just as a large standing army, with its small annual levies of serfs, had satisfied Russia's military and social needs for a century and a half between the Great Northern War and the Crimean War, so too had the nation's needs for weapons been met without great difficulty. A state-run arms industry based on the French model of royal monopoly in arms production made Russia virtually self-sufficient in arms and met the Russian government's obsession with national security, its mistrust of its subjects to supply arms in times of need, and its reluctance to pay for its great power needs. Moreover, the weapons of the day were relatively simple and, more important, technologically static, both in system design and in methods of production. Weapons did not have to be procured suddenly in large numbers; they could be stockpiled and, if given reasonable care and maintenance, reused for generations. To the mid-nineteenth century, and even beyond, as Walter Pintner argues, the greatest share by far of the budget went for subsistence expenditures such as pay, food, fodder, and uniforms. Only a small proportion—little more than 10 percent—went for weapons and ammunition.[7] The political and social system effectively mobilized the resources of a large but poor country; and if weapons and munitions were an insignificant part of military expenses, they did not have to be, given Russia's political and social system, its military strategy, and its needs of war.

However, the defeat at Crimea and the uncertainty over rapid technological changes in weaponry created a sense of urgency at the highest levels and stimulated interest in new weapons systems, new theories and tactics of warfare, and new manufacturing methods. Minister of War Dmitrii Miliutin, in his report of 1862 outlining plans for sweeping military reforms, starkly stated the lessons of Crimea: "In the present state of the art of war, technology has become extremely important. Improvements in weapons now give a decided advantage to the most advanced army. We became convinced of this truth by the bitter experience of the last war. Now we must honestly admit that we are materially behind other European countries in our munitions and weapons."[8] An Artillery Department circular was more blunt: "Russia cannot, and indeed, must not lag behind the other powers in the sweeping rearmament of its army whatever the cost to the state."[9] This "sweeping rearmament" came to Russia at a time of great domestic ferment. To be sure, the Great Reforms had less direct impact on technological modernization than on military service. In many ways technological change had a life of its own independent of political reform. The need for more advanced military technology was a part of the relationships forming between a modernizing military and society. As Alfred Rieber stated, "In the technological age it was not enough to maintain a great army in order to claim the title of a great power; indeed it was impossible without literate men, modern machines, and an efficient administration."[10]

The needs of nineteenth-century armies and rapid changes in weapons systems in Europe and America stimulated technological change, economies of scale, mechanization, and mass production. Centers of arms production had long been "islands of modernity" in a preindustrial world."[11] Yet while war provides a powerful motive for technological change, it cannot predict the occurrence or rate of diffusion of production innovations. Recent studies of innovation diffusion agree that the adoption of an innovation is dependent on research and development: effective information flow, knowledge of production needs, design organization, and assembly. The importance of such research and development in turn suggests that gradual, incremental, and unspectacular changes—what might be called "normal technology"—are often more important than spectacular inventions, or technological revolutions, in the diffusion of new technology, and that growth in productivity results not only from technological change but also from advances in the organization and management of existing technology. In the West, the diffusion of new technology through research and development has been facilitated by the interdependence of industries and of the military and civilian sectors of the economy.[12]

Paradigmatic of the above processes was the nineteenth-century American arms industry. Studies of American arms production have provided evidence of far-reaching changes in technological innovation and diffusion, the relationship between government and industry, the methods of working metals, and the organization of work.[13] The American model of industrial development during the first half of the nineteenth century provided many points of integration between military and civilian sectors and between government and private small arms production. By mid-century the result, in the phrase coined by a British parliamentary commission, was the "American system of manufactures," characteristic of the small arms industry in particular.[14]

Despite the availability of new technology, its diffusion is not always immediate. In a history of the Harpers Ferry Armory, Merritt Roe Smith puts it eloquently: "Few societies relish change. It taxes customs, unsettles social relations, intensifies anxieties, and upsets the general rhythm of life in a community. Change is even less welcome when it challenges people who are securely rooted in a familiar environment and satisfied with the way things are."[15] Among the most important constraints to the diffusion of new technologies are prohibitively high costs of an innovation; the absence of a demand—whether market ("from below") or administrative ("from above")—for innovation; geographical or sectoral isolation of the "islands of modernity"; deficiency of entrepreneurial talent and/or of a reservoir of manual skills; conservative, risk-averse behavior; and "technological indigestion," that is, the inability of the productive system to absorb quickly new technologies.[16]

Developing countries desiring a nation in arms but lacking a nation of armorers and mechanized industry have always been faced with a choice between costly arms purchases from foreign suppliers and even more costly arms production at home. This choice illustrates the dilemma of development: the nation urgently faces the need for technological change, yet it is precisely the resistance to change that makes the nation underdeveloped in the first place. A government that cannot easily produce arms in wartime perceives itself to be at best a hostage of rapacious producers, middlemen, and foreign suppliers, or at worst defenseless. For those nations following the path of domestic arms production, the goal has been industrial independence. As recipients of foreign technology, most developing countries present a "negotiated environment," a static environment protected from the free play of competitive forces and characterized by a niggardly utilization of capital resources, a scarcity of entrepreneurs, and an immobilization of potential human resources. Activist government agencies set the technological agenda, purchase foreign technology, and hire foreign technicians while jealously maintaining state supervision or control over the process.[17] A highly negotiated environment is the market for military arms. Once a developing country has made the decision to produce arms domestically and to generate the enormous amount of capital an arms industry requires, a rather predictable series of steps follows. The country sets up facilities to service and overhaul weapons; it obtains licenses to assemble parts produced in other countries; and, finally, it acquires the capability to produce an entire weapons system from design to manufacture. Yet technology transfer is not always readily absorbed and does not offer a panacea to the ills of development. The "absorption gap," the effective incorporation and assimilation of the foreign technology into the receiving country, is frequently longer than the actual transfer of technology, the "imitation lag."[18]

Russia was one of the first countries to face this dilemma of development. To begin with, technological change came from above (the state) or from the outside (foreigners) and was motivated by administrative considerations. Financial motivations did not dictate new technologies: there was little pressure for profitability from the marketplace. The many handicaps to private entrepreneurial activity, native and foreign alike, only made matters worse. Poorly developed capital markets, the necessity for government approval, the shortage of money, and weak domestic demand, especially for consumer goods, militated against domestic incorporation and created a highly negotiated environment for foreign entrepreneurs. Although foreign masters and supervisors could help introduce production patterns that native enterprises could then domesticate, foreign capital was regarded in many quarters as a threat to sovereignty, and foreign business activity was particularly closely supervised. This, coupled with the fact that the government was frequently the only consumer of foreign products or technology, effectively kept away most American businesses, which tended to have few contacts with host governments abroad. As a result, private enterprise, an energetic partner to governments in the development of native human and material resources, played only a small role in innovation and the diffusion of new technologies in Russia.[19]

The Russian small arms industry has "fallen through the cracks" in virtually all studies of nineteenth-century military and society, of economic development and technological change, and of industry and labor. Older studies of military and society have concentrated on the army under Nicholas I, histories of battles, military thought, or histories of weapons.[20] Histories of the military reforms of the 1860s and 1870s are heavily biased toward institutional change.[21] Newer works on military and society have concentrated on civil-military relations, the peasant in uniform and in mutiny, and tactical doctrine but, with few exceptions, have barely mentioned arms production.[22] Studies of the emancipation of the serfs understandably focus on the proprietary and state peasants but in the process neglect the indentured labor force in government factories, particularly in the armories.[23]

Studies of economic development and the role of the state in the economy have focused on the late nineteenth century, the role of Sergei Witte, and the private entrepreneurial elite.[24] Foreign business operations have found their historians, but the latter have neglected the small arms industry.[25] Although the venerable Soviet historian Mavrodin has pointed out that the development of firearms is an integral part of the development of technology and material culture,[26] it seems that few historians have followed his lead. Older studies of Russian industry and workers have focused largely on the textile industry; on heavy industry in St. Petersburg, the Urals, or the Donbass; or on aggregate production figures.[27] Studies of local crafts, particularly those of the Tula region, generally have omitted the armorers.[28] Although the world of management and labor has received increasing attention lately, and although the importance of skilled metal workers has been recognized, the small arms industry has largely remained unexplored.[29] The few Soviet studies of small arms and small arms production typically do not go beyond chronicles of weapons adoption, factory histories, or labor struggles.[30]

Following defeat in the Crimean War, the Ministry of War engaged in a massive rearmament program, "the most energetic measures," in Miliutin's modest words, to modernize both design and production of its small arms. Whereas the enormous expense delayed the adoption of new technologies and the supply of artillery pieces, one rifle model after another was adopted and supplied to the infantry. During the 1860s and 1870s, Russian officers made exhaustive studies of native and foreign systems, frequently visited foreign small arms factories, and undertook extensive tests of weapons purchased abroad. By the time of Russia's next major military engagement, the Russo-Turkish War, an army that had traditionally eschewed infantry firepower, relied on the bayonet charge in closed formation, and feared that soldiers would waste expensive ammunition had adopted breechloading rifles, revolvers, and metallic cartridges. The Russian military turned to a new source of innovations in firearms, and in so doing the Russian small arms industry began to interact with the American—the premier foreign small arms industry and the source of major advances in small arms design and production. The Imperial government dispatched ordnance experts to "the arms makers of the Connecticut valley" where they tested and purchased weapons, adapted American firearm and cartridge designs to Russian specifications, and supervised the manufacture of arms purchased by the Russian government.

The Crimean War exposed Russia's cumbersome military procurement from inefficient producers. Aware of the danger of dependence on foreign arms suppliers, the Ministry of War became more directly involved in the development of the arms industry than ever before. Unlike another developing country of the day, Turkey, but very much like the later Soviet government, Russia imported a limited number of modern arms while concentrating its efforts on the creation of a military-industrial establishment with the machinery and technological know-how to modernize domestic production and to achieve independence from foreign sources in wartime. Despite the claims of Soviet historians that "the Tsarist government, having superb gunmakers, let its slavish kowtowing to foreigners and its ignorance of the creative abilities of the Russian people dictate a preference for ordering weapons abroad, thereby subsidizing foreign businessmen,"[31] Russia in fact continued to produce most of its small arms domestically. In one of the most important transfers of arms technology in the nineteenth century, American manufacturers supplied the Russian government with machinery, machine tools, production organization, and even skilled workers.

But for a poor country modernizing its weapons, the cost factor was critical. In trying to keep up with the pace of technological change, the government operated from one basic premise: the need to supply large numbers of reliable arms to the state at the least possible cost under conditions of rapidly changing weapons systems and of freed, and therefore more costly, labor. In the process, the government ran into several constraints to technological change and innovation diffusion. These constraints can be best understood if we examine the adoption of new weapons in the context of labor-management relations, the reorganization of work processes in the small arms industry, the processes and costs of technological innovation, and the role of the state in the Russian small arms industry.

The manufacture of small arms placed a severe strain on the Russian economy. The strain of frequent model changes required quick adaptation, flexibility, and diffusion of new technologies. Moreover, the emancipation of the serfs altered the relationship between labor and management in factories that had employed a serf labor force. The transition between serf labor and free labor and between state management and private management was nowhere more difficult than in the government armories, especially in the largest and most important, the Tula armory. Although the Russian government spear-headed the modernization drive in small arms a generation before it spearheaded the better-known modernization drive of the 1890s, the industry was unable to reproduce native technology, let alone domesti-cate foreign innovations. By the end of the nineteenth century, the Russian government was again searching in foreign markets for the newest technologies in small arms. The debate generated by armory practice reveals much about the relationship between government, military, and industry, the attitude toward private entrepreneurship, and the stimulus to innovation in a developing country.

FIREARMS IN THE INDUSTRIAL AGE

● IN THE NINETEENTH CENTURY, THE long-service standing army gave way to the short-service conscripted army. This change affected military and society, forcing nations to upgrade their potential for war. Governments needed an ever larger number of armaments, trained soldiers, raw materials, and skilled workers. In the middle of the nineteenth century, three fundamental changes in weaponry and technology—the long-range breechloading infantry rifle, the mechanization of the arms industry, and the mass production of weapons—turned the mass army into an engine of destruction.[1] Military technology knew no national borders, and weapons designers and arms makers found eager customers. Between the Crimean and Franco-Prussian wars the rapidly changing weapons, tactics, arms production, arms marketing, and technology transfer in Europe and America defined great power status in the industrial world. When the ability of Russia's army and society to compete with the West was seriously called into question in mid-century, Russia found itself "up against" this new industrial matrix.

Evolution of Small Arms Systems

Firearms had been used in warfare as far back as the fourteenth century. Hand guns were first used in the fifteenth century, and the heavier firearm that came to be known as the musket was developed in the middle of the sixteenth century. Meanwhile, mechanical ignition systems progressed from the matchlock in the late fifteenth century, to the wheellock in the sixteenth century, and to the flintlock in the seventeenth century. The flintlock, invented in 1635 and in general use by the end of the 1600s, remained the standard ignition system of the infantry arm until the 1840s. Rifling was invented in the early sixteenth century, but, because it was difficult to combine muzzleloading with a bullet adapted to rifling, it took more than three hundred years for rifles to become widely used. The early sixteenth century saw the first breechloading hand guns and the later part of the century saw the first bullet and charge of powder in a paper envelope, or cartridge. But breechloading guns could not be used until precision metalworking made it possible to produce an airtight breech and a properly fitting cartridge. Although the innovations that were to figure prominently in the rapid advances in small arms in the nineteenth century—rifled bore, breechloading, and cartridges—had long been known, they were beyond the technical capacity of the day. The infantrymen of the Napoleonic Age knew only the smoothbore, muzzleloading, bayonetted flintlock musket, the "Brown Bess" of British lore.[2]

During the first half of the nineteenth century, several technical advances enabled a dramatic increase in firepower and quickly rendered the Brown Bess obsolete. The first important advances were the paper cartridge and the percussion cap. In 1807 a Scottish minister, Reverend Alexander Forsyth, patented a percussion powder for priming, and in 1814, Thomas Shaw of Philadelphia invented the percussion cap using a copper cap with fulminate priming powder that exploded by the blow of the falling hammer. Percussion ignition reduced the number of misfires, especially in wet weather, thereby increasing the effectiveness of the musket. Later it also facilitated breechloading. By 1840, the British army was equipped with percussion muskets. Nevertheless, other shortcomings of the smoothbore, muzzleloading musket—inaccuracy, limited range, and slowness of loading—remained. As late as mid-century, an American ordnance expert observed of the armies of Europe, "A large proportion of the infantry forces of most of the great powers of Europe are still armed with the ordinary musket, on account of the difficulty of making a general and sudden change in the arms of great bodies of men."[3]

Unless the bullet fitted the rifled bore tightly, it lost accuracy; yet a tight-fitting bullet was slow to load. Developments in rifling and in ammunition alleviated these shortcomings. Delvigne in France in 1826 and Greener in England in 1835 both made advances in expansive bullets that were easier to load. In 1849, a French army officer, Captain Claude Étienne Minié, patented an elongated bullet—first conoidal, later cylindro-conoidal—whose hollowed base expanded to fit the grooves of a rifled bore when the charge was fired. The combination of the Minié bullet, percussion cap, and the rifled bore produced an arm that required no ramming and was relatively easy to load, with increased accuracy and a longer range. This was the final improvement in the fire-power of muzzleloading arms.[4]

Despite advances in rifling and the invention of the percussion cap and Minié ball, the muzzleloader still had to be reloaded in a standing position. This took time—the rate of fire was still usually less than two rounds per minute—and was dangerous. A charge loaded at the breech, however, promised less cumbersome and safer reloading. The breech mechanism by which the gun is loaded, fired, and unloaded is called the "action." Perhaps the first breechloading arm to be exten-sively used was one patented by the American Colonel John Hall in 1811; originally a flintlock, it was later adapted to percussion ignition. In 1827 Johann Nikolaus von Dreyse produced a bullet containing a fulminate in its base and detonated by a needle; later he developed a bolt-action breechloader in which the trigger let fly a needle that penetrated a paper cartridge, setting off a detonator. In 1841 the Prus-sian army adopted Dreyse's breechloading rifle, called the "needle gun." The needle gun could fire seven shots per minute compared with two shots for the Minié muzzleloader. (Later muzzleloaders did have a longer range, and another breechloader, the smaller-caliber French Chassepot, was lighter, more accurate, and had a longer range.) The rapid rate of fire gave the Prussian soldier with the needle gun a considerable psychological advantage. In addition, the breechloader could be loaded and fired while the soldier was lying on the ground, was safely and easily cleaned, and had a simple mechanism.[5]

The needle gun was not without defects, however. A weakened nee-dle or spiral spring was apt to break or bend and thus not penetrate to the primer; the gun used a paper cartridge; a large amount of gas escaped at the breech; and a large amount of ammunition was ex-pended. Its production posed additional obstacles. Though smooth-bores were plentiful, it was very difficult to convert them to needle guns. Needle guns were precision instruments and costly to manufac-ture. Consequently, though the needle gun received much attention, it was not universally adopted. The American ordnance expert cited above wrote, "Prussia is the only large state in which an arm of this kind has been exclusively adopted for military service, and the absence of any imitators of her system in this respect seems to be a tacit ac-knowledgement of general disapproval of it in other countries."[6]

These words were written on the eve of a flood of improvements in breechloading systems—between 1860 and 1871, five hundred pat-ents on breechloading mechanisms were taken out in the United States alone. The rapidity of fire was increased by the "bolt action," a breechblock or breech bolt that inserted the ball and cartridge, sealed the breech, locked the cartridge or shell into firing position, and cocked the firing pin. An extractor on the bolt face removed the unfired cartridge or empty case from the chamber. "Single-shot action" extracted the fired cartridge and inserted a fresh cartridge in one movement. Between 1857 and 1861 four breechloading carbines—the Sharps, Terry, Green, and Westly-Richards—were experimentally introduced into the English cavalry. In 1864, England adopted a system designed by the American Jacob Snider for converting the muzzleloading Enfield rifle into a breechloader with a hinged breechblock. Later perfections in the single-shot breechloader were the Peabody and Martini-Henry rifles with falling breechblocks and the Berdan rifle with a sliding breechblock. In 1867 England adopted the Martini-Henry breechloader with an improved lock mechanism using a metallic cartridge. The Werndl and the Werder, variations of the Peabody, were similar models in Austria and Bavaria. In 1866 the French adopted the bolt-action Chassepot, which fired paper cartridges. Moreover, the substitution of steel for iron increased barrel endurance and reduced the number of explosions and misfires due to barrel defects.[7]

One of the drawbacks of the breechloaders was the escape of gas at the breech. Developments in cartridges eventually eliminated this problem. Samuel Johannes Pauly, a Swiss, invented the first cartridge containing its own primer, and in 1812 he patented a breechloading mechanism using a paper cartridge with a metal base. The percussion cap made possible the expansive cartridge case that permitted obturation, or the sealing of the breech to prevent escape of gas. However, such sealing required precision work in the breechlocking mechanism that was beyond the capability of the day's metallurgy. Metallic cartridges that contained built-in ignition, projectile or bullet, and expanding case provided the ideal gas seal. As in the needle system, the metallic cartridge could be fired without a capsule. However, in contrast to the needle system, the metallic cartridge had no spiral mechanism with a charge in it; instead, the charge was put directly in the metal case, thereby simplifying manufacture. Metallic cartridges consist of three types. In the pinfire cartridge, invented in 1836 by Lefaucheux in France and improved in 1847 by Greener in England, a pin projecting from the case near the base is in direct contact with a detonating compound. In the rimfire cartridge, invented by Flobert in France in 1847, the detonating compound is located on the rim of the cartridge base. In the center fire cartridge, invented by Pottet in France in 1855, the primer, or percussion cap, is fitted in a cavity at the center of the base, which yields a more positive ignition.[8]

Compared to the needle and percussion systems, the metallic cartridge was safer, more durable, more protected from moisture, and
more accurate; it also did away with cumbersome loading and powder-
carrying equipment. Although metallic cartridges had many advantages over paper—easier to use, fewer parts, more durable, supplied
in larger quantities—for a long time the high cost of and difficulty
in their manufacture prevented widespread use. Unlike paper cartridges, metallic cartridges could be made only in factories, not in
the regiments by the troops themselves. In addition, high quality copper was needed for the preferred alloy, brass, in cartridge manufacture.
For this reason, adoption of the metallic cartridge became possible
only when the metal industry was at a level sufficient to mass produce
them and when high-grade Lake Superior copper could be mined and
transported. For a country with a small standing army, such as the
United States, metallic cartridges were far superior to paper cartridges
for storage. Metallic cartridges made possible the success of, as well as
many of the subsequent improvements in, breechloading small arms.[9]

Although Switzerland and France provided the various cartridge systems, the United States administered the first practical application
of the systems. Daniel Smith and Horace Wesson developed the first
practical rim-fire metallic cartridge in 1856, and ten years later A.
C. Hobbs, of Union Metallic Cartridge Company, and Hiram Berdan
developed the first practical center-fire primer, consisting of a primer
cup, priming compound, and foil cover. Metallic cartridges first received extensive military use during the American Civil War. After
the Civil War, the news of the military advantages of metallic cartridges reached Europe, and a succession of foreign contracts kept
American factories busy.[10] As a correspondent for the London *Daily
Telegraph* lamented,

> The execution of such large orders by America—few seem to have come
> to England, our export of arms and ammunition having fallen off instead
> of increasing lately—affords another illustration of the facilities which their
> multiplied use of machinery give to the Americans. They are enabled to accept
> orders for such goods as firearms on a scale which our makers obviously
> cannot undertake; and the only explanation is that English contractors cannot
> engage to complete them by the time desired, while their transatlantic rivals
> can.[11]

Advances in metallic cartridges in turn paved the way for the deadly
weapons of the end of the nineteenth century. Various box- or tubular-
shaped devices for storing cartridges in the rifle and a feeding mechanism for delivering live cartridges into the chamber made possible
the magazine rifle. The Spencer and Winchester rifles were early ver-

sions of the magazine rifle, and by 1890 most European armies had adopted the magazine system. Soldiers could fire several shots without reloading manually after each shot. However, because of the increased rate of fire, magazine rifles exacerbated one drawback remaining in the metallic cartridges: the use of black powder. In volley fire, clouds of black powder smoke hindered the rifleman's vision. To make matters worse, riflemen opening fire revealed their position at once. In 1884 M. Vieille in France developed powder that burned more completely. This powder permitted longer and thinner projectiles, a reduction in bore size, and greater accuracy and velocity; more important, it was "smokeless."[12]

The final improvement in the breech mechanism was self-acting or automatic action. The first successful mechanical gun was patented by Richard Gatling in 1862. Six to ten barrels were set around an axis, and turning a manually operated crank or lever that loaded, cocked, fired, extracted, ejected, and reloaded caused the gun to fire repeatedly. The era of the hand-operated mechanical gun ended when between 1881 and 1883 Hiram Maxim developed the first fully automatic gun that fired cartridges or shells as long as the trigger was depressed and cartridges remained in the magazine. As self-acting or automatic mechanisms in industry came to be called machines, so too Maxim's automatic gun was soon better known as the machine gun. By the end of the 1880s the "deadly scorpion" of the battlefield had achieved the firing rate of six hundred rounds per minute.[13]

The revolver also received much attention in the nineteenth century. While the pistol contained a chamber permanently aligned with the bore, the revolver was based on the revolving breech, which dated from the sixteenth century, to enable the user to fire successive shots from the same weapon without reloading it. Cocking the revolver aligns the chamber in the cylinder with the bore, and firing expels the bullet from the cylinder into the bore. In early forms of the revolver, several barrels on a circular mounting rotated and fired successively by a matchlock, wheellock, or flintlock. This heavy and cumbersome construction was improved and streamlined with the addition of a revolving chambered breech that fired successive charges through a single barrel. Although Elisha Collier invented a mechanical means for rotating the cylinder in the early nineteenth century, the flintlock was ill suited for rapid, repeating fire.[14] Consequently, the revolver had not been militarily or commercially successful even though the principle of the weapon had been known for centuries.

A critical breakthrough came with the invention of percussion ignition and the replacement of the flintlock by the percussion cap in the early nineteenth century. In 1831 the first practical revolver mechanism employing percussion ignition was invented in the United States, a nation that, without a large standing army, had participated little in the European advances in infantry arms. The inventor, whose name was to become famous in the courts of Europe as well as on the frontier of America, was Samuel Colt.

In 1835 Colt obtained his first patents for the revolver from the English and French governments; one year later the American government issued to him a patent for a rotating cylinder containing six chambers, all of which discharged through one barrel. In the same year, Colt founded the Patent Fire Arms Manufacturing Company in Paterson, New Jersey, to begin construction of his revolver. He was not successful in getting the United States Army to adopt the revolver—it was considered too complicated—and the small number of orders forced him to close the Paterson plant in 1842. For the next five years Colt preoccupied himself with submarine explosive devices and his guns might never have been heard from again had it not been for the Mexican War. Having received an order from the government for one thousand revolvers, Colt resumed their manufacture near New Haven. He soon moved to Hartford, and in 1854 he began construction of what was to become the largest private armory in the world.[15]

Colt's original revolver initiated a flood of improvements and imitations. His most serious European competitors were Lefaucheux in France and Robert Adams in England. Like lesser-known American revolvers of the time, Colt's revolver was single action; that is, the hammer had to be cocked by the thumb to revolve the cylinder. English gunmakers effected this action by the pull of the trigger. Adams developed a modified version of the self-cocking mechanism whereby the trigger cocked the hammer and fired the charge in one movement. So-called double-action was achieved with the combination of single-action and self-cocking mechanisms. With this innovation, the hammer could be either thumb-cocked or cocked and released by pulling the trigger, which increased the rapidity of fire. Whether this also increased accuracy, range and penetration power was disputed. Controversial tests at the English arsenal at Woolich seemed to favor the Adams revolver. Although the Colt revolver was heavier, it was more durable.[16] The Adams revolver could not compete with either the publicity or the mass production of Colt. An article in the standard English military journal noted, "No one has contributed so much to alter the nature of warfare and the fashion of our arms as the American colonel. He it was who first gave us the revolver, and it is he who has since perfected the principle, and applied it to every description of fire-arms."[17]

After the Civil War, Colt's revolver models, still using paper cartridges and separate percussions caps, became technically obsolete. Moreover, after the expiration of his patent monopoly, Colt failed to deal with increasing competition. The deaths of Colt in 1862 and of his chief engineer, Elihu Root, in 1865 deprived the company of managerial talent. Although the company continued to promote the revolver aggressively in foreign markets, the relatively outmoded Colt revolvers did not sell well. According to the European agent of one of Colt's competitors, the single-action powder and ball pistols were unsaleable in Europe.[18]

The greatest threat came from Smith and Wesson, a company established in 1857. In 1856 Horace Smith (1808–1893) and Daniel Baird Wesson (1825–1906) produced the first practical self-contained metallic cartridge and the revolver that fired it. Although no revolver using metallic cartridges had been adopted at the time of the Civil War, the advantages were so apparent that thousands were bought by army and navy officers. In addition, Smith and Wesson developed the center fire cartridge, an advancement over the rim fire and pin fire cartridges. According to a contemporary authority on small arms, the .22 and the .32 caliber revolvers accomplished "far more than any revolver before invented" and were more automatic than any previous revolver. During the 1860s, the Smith and Wesson metallic cartridge revolver made serious inroads into the Colt business, and in 1869 Smith and Wesson sold $180,000 worth of revolvers, one-third more than did Colt. At the end of the Civil War, demand was so great for the .22 and .32 caliber revolvers that Smith and Wesson had orders for two years in advance.[19]

With the exception of the percussion lock, few of the innovations after the Napoleonic Wars reached the soldier, and, except in Prussia, by mid-century the standard infantry arm was still the smoothbore, muzzleloading musket firing a round bullet. Armies were just beginning to adopt the rifle, and the revolver was still too new a weapon for practical military use. Until the middle of the nineteenth century, rifles were expensive and slow to load; therefore only select riflemen were equipped.

Weapons and Tactics

The evolution of nineteenth-century infantry tactics illustrates well the relationship between arms and the army. The tactics of the first half of the nineteenth century were designed to maximize the potential of the smoothbore musket. Infantry moved in columns and lined up in closed battle formation suitable for volley fire or in roughly square formation for protection from bayonet or cavalry charge. Eventually, special units of skirmishers carried rapidly loading needle guns or long-range Minié rifles. However, skirmishers in extended order merely paved the way for the assault of infantry columns and the engagement of "cold steel."[20]

The gradual advances in firepower brought by the needle gun, the Minié rifle, and breechloaders created an uneasy tactical balance between firepower and the bayonet charge; between infantry, cavalry, and artillery; and between offense and defense. The rapid deployment of mass produced weapons further complicated tactical rules and infantry drill.[21] The advantages and disadvantages of breechloading rifles were a case in point. Initially, breechloaders gave an advantage to the defense, which could inundate front lines with fire. In 1866 the Prussians using breechloaders outfired the Austrians using muzzleloaders by a ratio of six to one. Neither infantry nor cavalry could attack frontally against breechloaders. Consequently, in order for the offense to take advantage of the new weapons, certain changes in tactics were necessary. In the face of rapid fire from defenders using breechloaders, flexible formations and dispersed order replaced closed formation. In addition, because frontal assaults against infantry were highly dangerous (though they persisted in the Civil War, for example), maneuver and flank attacks became more important. The introduction of the rifle was particularly valuable to skirmishers and light infantry, and proponents of breechloaders, such as Wilhelm von Ploennies and Caesar Rüstow in Germany, claimed that the devastation of enemy morale inflicted by rapid fire would make the new weapon decisive in the hands of well-trained troops. To put it differently, increased firepower would have a greater effect on the enemy than the most daring bayonet attacks. At first deployed chiefly to prepare the attack, skirmishers more and more, as happened in the Franco-Prussian War, became the main combat forces. This, in turn, placed greater premium on subordinate commanders, individual initiative and intelligence, speed, and agility. Since infantry could hold its own against cavalry and since an attack could be repelled at longer ranges with greater accuracy, mounted cavalry operations became limited; cavalrymen became more effective by dismounting and using firepower.[22] Moreover, artillery was a less effective defense against the new rifles because sharpshooting riflemen could reach gunners. According to one British officer, "In truth, within effective range, breechloading rifles are infinitely more destructive than artillery. . . . Now artillerymen even at 1,000 yards are exposed to an accurate and rapid fire from the enemy's infantry, to which the only answer possible is the shell."[23]

Although the breechloader promised a greater speed and ease in reloading, a combination of technical and tactical considerations delayed the military use of this new weapon, and commanders were reluctant to accept the supremacy of fire. For a long time the inability to seal the breech to prevent the escape of gas lowered velocity, lessened the potential range, and reduced accuracy; early breechloaders were ineffective at ranges over seven hundred yards. Black powder caused fouling that hindered cartridge extraction and reduced the effective rate of fire. Moreover, the breechloader was considered too complicated for any army whose rank and file were illiterate; only experienced men could be trusted in skirmish lines. Finally, commanders feared that infantrymen would waste ammunition; empty cartridge boxes would be impossible to replenish during battle.[24]

Even though the United States led the advances in breechloading rifles, criticism of the new weapon was expressed in the United States Army. The chief of ordnance questioned "whether this facility of firing will not occasion an extravagant use and waste of ammunition"; likewise, the ordnance board reviewing John Hall's breechloading rifle in 1836 remarked that "an arm which is complicated in its mechanism and arrangement deranges and perplexes the soldier."[25] Studies of European armies at mid-century suggest that the high cost and difficulties in mass-producing new weapons, the concern that models tested might be superseded by improved models, and the reluctance to expose the security of the nation to untried weapons delayed universal adoption of the breechloader. Most armies in mid-century continued to place a high value on the bayonet charge. Even after the Crimean War it was argued that fire was only a preliminary to the decisive attack of the mass, that firepower remained subordinate to the shock attack, and that the venerable shock actions of the bayonet charge and of the cavalry still contained tactical value. The superiority of morale, training, dicipline, and dash was still not seriously questioned. Ironically, while breechloader against muzzleloader may have highlighted the advantages of firepower, as the former was increasingly introduced, the stalemate of breechloader against breechloader revived arguments for cold steel and morale.[26]

Although tactical doctrines emphasizing cold steel coexisted with newer doctrines emphasizing firepower and although wars of the mid-century contained their share of suicidal cavalry charges, breechloaders with effective ranges of one thousand yards made the American Civil War, the Austro-Prussian War, and the Franco-Prussian War rifle wars. In turn, by the close of the nineteenth century, what the rifle had done to the bayonet, the machine gun did to the rifle: infantry could no longer conquer a position defended by machine gun fire. The machine gun, coupled with the spade and barbed wire, launched an era of trench warfare and again gave infantry enormous advantages in defense. It also put direction back into the hands of regimental commanders and offset the lack of control and erratic firing attendant with dispersed formations. Moreover, as J. F. C. Fuller put it, with the advent of smokeless powder, "the old terror of the visible foe gave way to the paralyzing sensation of advancing on an invisible one, which fostered the suspicion that the enemy was everywhere."[27] The machine gun "had not so much disciplined the art of killing . . . as mechanized or industrialized it."[28]

The Rationalization of Arms Production

But it is more in the making than in the pattern of the gun that American manufacturers have revolutionized old systems, and marked out the method by which firearms are now produced far better and cheaper than formerly.[29]

The technical and mechanical innovations in firearms in the first half of the nineteenth century were accompanied by equally important innovations in their fabrication—changes that were essential for their widespread adoption later in the century. Over time, precision machinery replaced hand tools, and centralized factory production replaced decentralized craft production. By mid-century the result, in the phrase coined by a British parliamentary commission, was the "American system of manufactures." The salient features of the American system in general and of the small arms industry in particular were division of labor and specialization of function, uniformity of work method, centralization of the workplace, hierarchical chain of command, mechanization of routine and standardized tasks, precision measurement, sequential operation of machine tools, and uniformity or even interchangeability of parts.[30] Some features of the American system did not actually originate in America, and not all features were developed simultaneously. But after the Crystal Palace Exhibit of 1851, the features of American firearms seemed sufficiently distinctive to British arms experts to warrant creation of a parliamentary commission to tour American factories.

Government and Innovation in the Arms Industry

In the beginning both private gunmakers and governments produced firearms. In western Europe, the best-known concentrations of gunmakers were in St. Étienne, Liège, and Birmingham. The Liègois, in particular, had a reputation for settling all over Europe, in a "constant movement of experts and technicians to countries where the skill was needed."[31] However, beginning in the sixteenth and seventeenth centuries, armies were becoming the ultimate users and governments were becoming the purchasers of weapons. This resulted in less mobility for gunmakers and in greater government intervention in the gun trade. Two models of government control over the production of military arms evolved—government purchase from private contractors and government production monopoly.

In England, the government obtained firearms through an elaborate, and often unreliable, contracting system with private Birmingham gunmakers. Certain artisans made parts of the gun—barrels or stocks, for example—which were examined by government inspectors and then sent to other artisans for assembly. This system left considerable control over operations in the hands of the individual gunmakers; at the same time, fluctuating government orders created instability of employment. Although neither gunmakers nor government were completely satisfied with this contract arrangement, it suited the erratic nature of demand for military arms and the absence in England of a large standing army. Similarly, in Austria gun parts produced by private gunmakers were sent to the government arsenal in Vienna for assembly, and in the German states, private manufacturers made weapons under government contracts supervised by government inspectors.[32]

The experience of France provides an example of the second model of government control over the production of military arms. Beginning in the seventeenth century, weapons and munitions were manufactured under royal monopoly. As weapons designs and fabrication became increasingly complex and costly, European governments began taking more control over the arms industry. According to a study of military and society in Germany, with the introduction of the needle gun the Prussian Ministry of War gradually took over private factories and shut down or consolidated smaller plants. "The days of the individual gunsmith and the workshop type of arsenal were over and the Prussian government had entered the arms business on a massive scale."[33]

Whether military arms were manufactured under royal monopoly or contracted to private manufacturers, actual production of the flintlock and percussion muskets was slow and laborious. But since there were few changes in firearms during the era of the smoothbore muzzle-loading muskets, there were also few changes in their production. Accordingly, arms stored in royal arsenals for decades could be issued again and again to troops. Since stockpiled weapons could equip an army in the field, rapid production was rarely necessary.[34]

For a long time the English government was dependent for arms on private contractors; the American government at its birth was dependent on foreign sources. A shortage of gunmakers in America meant that many weapons were assembled from English and French parts. During the War of Independence, France, understandably, was the most important source of arms, and approximately eighty thousand muskets were imported. But American policy in the procurement of military arms deviated from the English. So as not to remain dependent on foreign producers, the new government decided to follow in part the French model of making arms in government armories and accordingly founded armories at Springfield and Harpers Ferry in 1794 and 1798, respectively.[35]

Despite the existence of two federal armories, no government monopoly on military arms foreclosed private production. Close cooperation between government and private industry was an important feature of the American model in arms production. Because at first facilities at these new armories were limited, when necessary the government still turned to private gunmakers to fulfill additional orders. However, dependence on the vicissitudes of the private arms trade was no better than dependence on foreign suppliers. Accordingly, the American government stepped in to help develop and to regularize the private production of arms. Although it has become fashionable of late in the United States to associate government intervention with economic stagnation and suppression of individual entrepreneurial activity, many of the largest American arms makers depended from the beginning on the government for contracts and patronage.[36]

To discourage the use of foreign components and at the same time to encourage a native arms industry, the war department favored contracting American gunmakers for complete arms. In addition, the government provided financial advances that constituted a major source of capital for private contractors, assured continuity of supply, and encouraged the building of manufacturing plants. The strict requirements for materials and quality dictated by the military fostered the rapid development of new products, processes, and tools. In his masterful study of Harpers Ferry Armory, Merritt Roe Smith argues,

This indeed was one of the most enduring contributions made by the government to industry during the antebellum period. Under the auspices of the Ordnance Department successive administrations encouraged, supported, and rewarded arms makers in developing a large staple of ingenious woodworking and metalworking machinery. These devices formed the keystone of modern interchangeable manufacture in the nineteenth century.

As a result of its patronage in the small arms industry, the government bore the high costs of mass production and machinery and thereby sponsored the industrial system of manufacturing.[37]

Many factors hastened the spread of new technologies. "Yankee enterprise" as embodied by Roswell Lee, the superintendent of the Springfield armory, was driven to mold men and machines into efficient systems. Military and civilian sectors of the economy were not segregated, and this sectoral integration coupled with the geographical proximity of the New England arms makers and the openness of the society in which they produced arms explains the quick flow of information and the assimilation of new technology. Geographical proximity to other manufacturing centers favored particularly the Springfield armory, which frequently transferred men, machinery, and methods to Harpers Ferry and to private arms makers in order to diffuse new techniques. By virtue of what the economist Nathan Rosenberg has termed "technological convergence," the machine tool industry and precision measurement in firearms brought together manufacturers and designers and provided configurations of machine-tool design that became standardized and easily diffused from one machine-using industry to another.[38] "Such interindustry flow of technology," according to Rosenberg, "is one of the most distinctive characteristics of advanced industrial societies."[39] Specialist, capital goods firms supplying a variety of industries have generated much of the technological change of the past two hundred years, and many of the benefits of increased productivity flowing from an innovation have been captured in industries other than the one in which the innovation has been made. Many spin-offs from the arms industry have benefited the civilian sector, and the debt of the bicycle, sewing machine, and agricultural implements industries to the firearms industry is by now well known.[40]

Nevertheless, despite the availability of new technology, as Paul Uselding and Merritt Roe Smith argue, its diffusion was delayed, even in the United States. "Technological indigestion," the inability of the productive system to absorb quickly the new technologies (in this case, machine tools) caused a time lag between the introduction of precision machinery and the full impact of production measures. Just as proximity to other producers and consumers and interdependence of sectors stimulate innovation by providing effective information flow, their absence acts as a constraint. Geographical isolation was central in the development of Harpers Ferry: "If one factor conditioned Harpers Ferry's attitude toward change, it was the lack of contact with the outside world. Unlike Springfield which was part of a rapidly growing

industrial region, Harpers Ferry remained secluded in a sparsely popu-
lated agrarian hinterland."[41] Segregation of designers from manufacturers and the subordination of civilian to military producers have been
the bane of many arms industries. For example, while the French
model of royal monopoly in the production of small arms allowed
arms producers to enjoy a seller's market, there was no pressure to compete or to change; at the same time, it accorded no control to the private
sector. As a result of this subordination of the civilian sector, according
to François Crouzet, France alternated between technological advancement and technological backwardness, incapable of utilizing and diffus-
ing earlier innovations.[42] Finally, individual constraints have frequently
complemented institutional constraints to innovation. Conservative,
risk-adverse behavior and the fear of loss among manufacturers and
workers alike have constrained innovation and the diffusion of new
technology.

Men and Machines

For centuries, the gun trade had been organized along craft lines,
with all the preindustrial culture of labor that the craft system entailed. From the beginning, the complexity and high cost of firearms
made gunmaking a highly skilled craft. Individual artisans made the
component parts, the intricacy of which necessitated skill and time.
Locks, of course, required intricate forging, grinding, filing, and finishing; likewise, adjusting and fitting component parts to make a finished
product—for example, fitting and recessing the gunstock to accommo-
date the lock and the barrel—required a great degree of costly skill
and patience.[43] The high unit costs of artisan-intensive manufacturing—what the English described as the "handicraft plan"—
were tolerable for the individual, wealthy users of firearms. Like other
skilled crafts, gunmaking was an art passed from master to apprentice.
The gunmaker's craft was a calling, a way of life, not merely a way
of making a living. The gunmaker was a respected member of the
community; paternalism, patronage, and deference set the tone of
community relations in the preindustrial world.[44]

A rudimentary division of labor in the gun trade had existed for
a long time. But in the United States the shortage of skilled gunmakers
forced factory managers to simplify production procedures. This was
accomplished by greatly subdividing labor tasks and by introducing
machinery. Government contract policy facilitated both. At the Harpers Ferry armory, the division of labor introduced in 1809 resulted,
by 1816, in fifty-five different operations (versus twenty in 1810) in
musket making; lock-making alone changed from two separate opera-

tions to twenty-one. Altogether, the number of occupational specialties jumped from thirty-four in 1815 to eighty-six in 1820 and one hundred in 1825.

Along with the division of labor, piece-rate accounting was also introduced, first at the Springfield armory. Piece-rate accounting was actually the application of an old procedure, the putting-out system, to a new setting, the factory. The new accounting system "recorded the product of each gunmaker by job and measured the output in terms of individual operations performed on a single component rather than on complete assemblies."[45] By 1810 the piece-rate practice had spread from items contracted outside the factory to small components manufactured within the Springfield armory. Moreover, the division of labor, piece-rate accounting, and the system of inside contracting made the position of master armorer, chief engineer, or superintendent even more critical in the coordination of output. According to one student of the Springfield armory, management used these changes as a method of control over labor and material.[46]

The division of labor, as Adam Smith teaches, promotes innovation. The most readily apparent innovation was the introduction of gunmaking machinery, which has figured prominently in contemporary accounts and in histories of technology.[47] The history of the application of machinery in the arms industry is a combination of three intertwined histories: the invention and application of a machine to make a single component or to solve a single production task, the use of precision machine tools to make uniform or interchangeable parts, and the sequential operation of special-purpose machines to facilitate manufacturing in a factory setting. The mechanization of firearm production during the first half of the nineteenth century provides examples from all three histories.

The leaders in mechanized production generally were England and the United States, and to a lesser extent Germany. The pioneering example most often cited by students of the history of technology is the blockmaking machinery invented by Marc Brunel and Henry Maudsley (the latter called the "father of the English machine-tool industry") and promoted by Jeremy Bentham. Used at the Portsmouth naval yard in 1808, forty-four machines made it possible to replace one hundred skilled workers with ten unskilled workers. Other examples of mechanized production in England include the "self-acting" mule of Sharp and Roberts patented in 1825 and the "self-acting machines" of James Nasmyth's Bridgewater Foundry established in 1836.[48]

An important impetus to the mechanization of production was the introduction of the Prussian needle gun. A precision instrument, the needle gun required exact fitting and standardization, which could only be achieved by machinery.[49] On this side of the Atlantic important machines used to solve specific production problems included the first gun-stocking machine in 1818, the die-forging machines in 1827, the first power rifling machine in 1832, the sets of gauges and jigs used at both the Springfield armory and the Ames Manufacturing Company in 1842, the turret lathe introduced in 1845, and the first universal milling machine in 1852. The adoption of metal-shaping technology made possible by the turret lathe and milling machine put the firearms industry far ahead of other branches of manufacturing and provided a reservoir of skills and technical knowledge. According to a contemporary expert on manufactures, the milling machines were the most numerous and characteristic of gunmaking machines and were applied to a great variety of jobs. During the second half of the nineteenth century, approximately one-quarter of the machines of a typical small arms factory were milling machines.[50]

As recent histories of American technology have shown, the labor-saving woodworking machines were resource-intensive and made possible because of the abundant and cheap supply of wood, a factor with which English manufacturing was less favorably endowed. The pioneer in the development, manufacture, and sales of woodworking machinery, especially for the fabrication of firearms, was the Ames Manufacturing Company of Chicopee, Massachusetts. Abundance in another resource, high-grade Lake Superior copper, and the development of mining capabilities and transport networks to ship copper to eastern processing centers permitted mechanization in the production of metallic cartridges, the crucial component of breechloading rifles. High-quality copper mined in the Lake Superior region was then semi-finished into sheets, rolls, and wires of brass by Coe Brass Manufacturing Company of Walcottsville, Connecticut, and delivered to cartridge manufacturers such as Union Metallic Cartridge Company of Bridgeport, Connecticut.[51]

The transition from craft to machine production of guns changed the concept and organization of work. As mechanization increased the precision and speed of metalworking, it increased the constancy and intensity of work. The workplace gradually became more disciplined, time-oriented, specialized, and centrally controlled; control inexorably slipped from the hands of the skilled craftsmen. In the candid words of a nineteenth-century expert on manufactures, machines have "rendered the government in no small degree independent of the skill and power of workmen."[52] In turn, mechanization facilitated two other innovations in production pioneered in the American firearms industry—development of interchangeability and the spread of the factory system.

Uniformity and Interchangeability

The so-called uniformity system actually originated in France. In 1765 General Jean-Baptiste de Gribeauval "sought to rationalize French armaments by introducing standardized weapons with standardized parts." At the same time manufacturer Honoré Blanc tried to achieve uniformity in musket parts, despite the clumsy wooden machinery then available. In a letter to John Jay in 1785, Thomas Jefferson suggested introducing his idea into the United States:

> An improvement is made here in the construction of the musket which it may be interesting to Congress to know, should they at any time propose to procure any. It consists in the making every part of them so exactly alike that what belongs to any one may be used for every other musket in the magazine. The government here has examined and approved the method, and is establishing a large manufactory for this purpose. As yet the inventor [Honoré Blanc] has only completed the lock of the musket on this plan. He will proceed immediately to have the barrel, stock and other parts executed in the same way. Supposing it might be useful to the U.S., I went to the workman. He presented me with the parts of 50 locks taken to pieces at hazard as they came to hand, and they fitted in the most perfect manner. The advantages of this, when arms need repair, are evident.[53]

Jefferson hinted at the principal motivation behind uniformity. A half century later a letter to the *Times* by Lieutenant Colonel W. M. Dixon of the Royal Army comparing the Enfield armory and the Birmingham gunmakers observed that interchangeable manufacture of the former promised to simplify repair and maintenance, so that new parts could be more easily and more cheaply substituted in the field:

> The question at issue is not merely the making of a rifle but that a rifle so made should resemble any and every other made in the establishment so completely that all similar parts should be capable of interchanging. If a soldier should lose his bayonet or ramrod . . . he may get similar articles to replace them at once without any special fitting being required. All parts of the arm are made on this principle and it can easily be conceived how valuable such a principle must be in an army which has to be armed entirely with one pattern of arm. Instead of an armourer having to forge, file, and specially fit any limb to a lock, finished parts will be supplied to regiments and will replace in a moment those which have been injured.[54]

In the United States, too, although seemingly far from Europe's battle-fields, military motivations were also paramount. According to Merritt Roe Smith, the primary stimulus for uniformity in the United States came from the army, "particularly its desire for more uniform and precisely made weapons whose components could be exchanged in the field for new ones whenever the need arose."[55]

Uniformity in production by machine, of course, required certain factors of demand and resource endowment. According to Nathan Rosenberg, the United States possessed a rapidly growing population, a high rate of new household formation, an egalitarian social structure, a weak craft tradition, and an abundance of natural resources. Uniformity in production also presupposed a willingness on the part of the public to accept a homogeneous final product. Such a willingness was absent in the British gun market, long dominated by taste and craft traditions that resisted uniformity. For example, the length, bend, and casting of the gunstock had to be fitted to the individual English user.[56] Colt himself, before the Parliamentary Select Committee, testified that the lack of uniformity was the major fault found with British firearms. Of the less varied American market for firearms, Colt stated, "It is the uniformity of the work that is wanted."[57]

Uniformity necessitated precision measurement, another characteristic of the American system of manufactures. According to Uselding, what distinguished the American system was the use of special-purpose machinery, "sufficiently accurate so that parts produced by machinery would be so close in dimensional tolerance as to permit assembly with a minimum of time for fitting."[58] The extensive use of gauges and the rationalized design of fixtures, first developed by John Hall at the Harpers Ferry armory in the 1820s, reduced the errors resulting from frequent "fixing" of an object in a machine tool. The universal milling machine in particular "displaced the highly expensive operations of hand filing and chiseling of parts" and permitted a high degree of uniformity. According to Rosenberg, the firearms industry was "instrumental in the development of a whole array of tools and accessories upon which the production of precision metal parts was dependent."[59]

The combined effect of the use of machinery and tools at the large armories was the establishment of a system of manufacture of interchangeable parts, the most striking feature of the American system. Interchangeability may be defined as a system of "mechanisms possessing closely fitting and interacting components in such a way that a given component of any mechanism would fit and perform equally well without adjustments in any of the other mechanisms."[60] The chief ingredients were the use of jigs for holding the work and guiding the tools, accurate machine tools, and gauges for frequent checking during manufacture.[61]

Legend has it that Eli Whitney introduced interchangeability into the American small arms industry and in 1801, before the secretary of war and a group of army officers, demonstrated that parts of ten muskets could be assembled at random. Recent research suggests that, though he used machine tools and factory methods, Whitney's muskets had merely more uniform, not perfectly interchangeable, parts. His government contracts, accordingly, resulted from his entrepreneurial and promotional rather than his technical abilities.[62]

More systematic and successful at achieving interchangeability was another private contractor, Simeon North. An 1813 contract from the war department for twenty thousand pistols specified that North make "the component parts . . . to correspond so exactly that any limb or part of one pistol may be fitted to any other pistol of the 20,000."[63] Because government armories had more time and money to pursue the development of interchangeability than did private companies, Springfield and Harpers Ferry made the most significant advances in this technology. For example, the use of inspection gauges introduced from 1817 to 1823 at the Springfield armory by Superintendent Roswell Lee was an important step toward interchangeability. Similarly, John Hall introduced precision instrumentation and uniform standards of measurement at Harpers Ferry; his 1819 breechloader used sixty-three gauges. By integrating lathes, drill presses, and other machines into production, Harpers Ferry produced the first truly interchangeable weapons in the United States.[64]

In 1853 the British Parliamentary Select Committee on Small Arms visited American government and private armories and recommended that England set up a government armory at Enfield modeled on the American system. In London the next year, the committee heard testimony from a variety of witnesses, including Colt, a great promoter of interchangeable manufacture. The committee, testing interchangeability, dismantled Springfield muskets, mixed up all the parts, and then reassembled the muskets from random parts.[65] Such demonstrations were a vivid testimony of the military significance of interchangeability. The committee questioned witnesses closely on machine production and interchangeability, the most striking claims of the American gunmakers. Joseph Whitworth, a Manchester tool manufacturer, stated that perfect interchangeability was impossible and that a machine-made rifle had to be finished by hand. Gage Stickney, the former superintendent at Colt's London armory, observed that each part of the revolver was finished by skilled labor and concurred that it was impossible to make a firearm entirely by machine. The inventor

James Nasmyth admitted that Colt did not put parts together directly
from the machine but "passes them through what they call the finishing shop."[66]

Colt, though proudly asserting, "There is nothing that cannot be
produced by machinery," did not go so far as to claim perfect interchangeability. After the Crystal Palace Exhibition, Colt noted, "In
Hartford the separate parts travel independently through the factory
arriving in almost complete condition in the hands of the finishing
workmen."[67] Three years later, in front of the Parliamentary Select
Committee on Small Arms, Colt claimed that his revolvers "would
do a great deal better than any arms made by hand. . . . In my own
arms one part corresponds with another very nearly."[68] Armory workmen filed and fitted machine-made parts while soft. When assembled,
the major components were stamped with serial numbers, the firearms
were taken apart, and the parts were hardened. Hounshell and Howard
argue that though Colt did employ a jig-and-fixture system, a gauging
system, and the sequential operation of special-purpose machines,
elimination of hand-fitting of parts was too costly for private arms makers.[69] The cost efficiencies of the new technology, easy to see with hind-
sight, were not evident to most manufacturers until the middle of the
nineteenth century. Although interchangeability promised a higher degree of specialization and a reduction, if not an elimination, of costly
fitting operations, Smith argues that initially machine processes introduced only marginal savings, and purely economic considerations had
little to do with the primary development and application of machinery.
The demand for uniformity came not from the economic calculations
of private arms manufacturers but from the military calculations of
the ordnance department and the national armories it supervised.[70]

Chapter 3 will show that claims of interchangeability were also
made in Russian armories; at this point it might be well to keep in
mind that the concept of interchangeability was relative, changing
over time. Smith points out that for the army, "the concept of uniformity [initially] meant little more than the degree of similitude between finished components that would allow field armorers to repair
damaged muskets from salvaged parts with a minimal amount of filing
and fitting."[71] According to Charles Fitch, a contemporary expert on
American manufactures, in the early nineteenth century, locks were
assembled "soft," marked, and then hardened. Filing and fitting were
the principal means of making interchangeable parts, and inspections
were not "severe." Consequently, joints were frequently uneven. By
1880 interchangeability signified "slipping in a piece, turning a screwdriver and having a close, even fit." In this sense of the word, interchangeability was not achieved until the introduction of "metalwork-

ing with sufficient machinery for making sensibly exact cuts, *without dependence on the craft of the operative.*" Close forging with steel dies enabled lock parts to be "much more exact than the *so-called interchangeable work* of early days."[72]

The Factory System

The small workshops of individual gunmakers could not achieve mechanization and mass production of uniform component parts. It will be remembered that the American government arms procurement policy differed from the British in the establishment of government armories and in the promotion of large-scale factory production through financial advances. Such support facilitated considerable economies of scale. To reap the advantages of large-scale production, machines were installed in a continuous flow arrangement. Production capacity at a single plant could be devoted to an individual product, made on long production runs. This necessitated the use of special-purpose equipment, jigs, and fixtures, extremely costly because they were job specific.[73]

Innovation in the factory setting, as well as the application of the sequential operation of special-purpose machines, was encouraged by the system of inside contracting. In this system, manufacturers and government armories provided shop space, power, machinery, tools, and raw materials—on the basis of competitive bidding—to subcontractors to produce a particular part or to manage a particular operation for a set piece rate. Receiving a foreman's wage, the contractor was responsible for hiring, training, managing, and paying his labor force. This inside contracting system, practiced at the Remington, Springfield, Robbins and Lawrence, Ames, and, perhaps most extensively, Colt factories, was process rather than product oriented; the subcontractors contributed improvements mainly in the productive process rather than in the final product itself.[74] As a result, according to a government report on manufactures, the workmen were "brought to take a more direct and active interest in the prosperity of the firm than is usually the case in large manufacturing establishments."[75]

That workmen were "interested in the prosperity of the firm" suggests the manufacturers' and engineers' ideal of a new culture of industrial labor. Indeed, government and contract policy, mechanization, and the establishment of complete manufacturing plants all undermined the culture of preindustrial labor. Task-oriented work was rhythmic, integrated with family or agricultural life, subject to few managerial constraints, tolerant of traditional observances such as drinking and gambling, not rigorously organized, disciplined, and centralized.

Possession of skill was a way of life, and a craft skill was an art, a "mystery," passed on by example.[76] It took entrepreneurs of "Napoleonic nerve," the apt phrase of a proponent of the factory system in England, to "subdue the refractory tempers of work-people"; to introduce a systematic work regimen, continuous employment of the mechanized factory, successive divisions of labor, inside contracting, piece-rate accounting; and to prohibit alcohol in the shops.[77] The Springfield armory, for example, was by 1820 no longer "a disorganized collection of craft workers" but "a disciplined assemblage of industrial workers."

> Had the labor force, then largely drawn from agriculture and accustomed only to the discipline of the seasons, been permitted to retain its old habits and attitudes toward work, it is safe to say that the innovations at Springfield associated with the development of interchangeable manufacture would not have taken place. . . . The "house-breaking" of the labor force was an absolutely necessary pre-condition for successfult innovation in the technical aspects of arms production that came to be known as the American system.[78]

At the Harpers Ferry armory the "transition from craft to machine followed a long and circuitous path" as labor and management alike, accustomed to a preindustrial culture of craft and community, resisted systematic work discipline, mechanization, and, ultimately, technological innovation.[79] In the private sector, Colt mechanized his factories in Hartford and London not simply to maximize output, lower costs, or achieve interchangeability. In a revealing moment his patent agent in London, A. V. Newton, pointed out in 1854 that Colt's aim was "not merely to make all the like pieces counterparts of each other but also to banish, as far as may be, the file from the workshop and thus while greatly expediting and reducing the cost of the manufacturer to render himself perfectly free of all combinations of skilled workmen."[80]

The result of Colt's endeavors was perhaps the epitome of the well-ordered factory. Certainly it seemed so to members of the visiting Parliamentary Select Committee on Small Arms. Nasmyth had been unimpressed with a visit to the Royal Arsenal at Woolwich in 1847:

> The mechanical arrangements, the machine tools, and other appliances were found insufficient for the economical production of the apparatus of modern warfare. . . . I made a careful survey of all the workshops; and although machinery was interesting, as an example of old and primitive methods of producing war material, I found that it was better for a museum of technical antiquity than for practical use in these days of rapid mechanical progress.

Everything was certainly very far behind arrangements which I had observed in foreign arsenals.[81]

Six years later Nasmyth's visit to Colt's factory left a very different impression:

It produced a very impressive effect, such as I shall never forget. The first impression was to humble me considerably. I was in a manner introduced to such a masterly extension of what I know to be correct principles, but extended in so masterly and wholesale a manner, as made me feel that we were very far behind in carrying out what we knew to be good principles.[82]

Nasmyth was particularly impressed with American tools, and his remarks aptly suggested the relationship between technology and culture:

In those American tools there is a common-sense way of going to the point at once, that I was quite struck with; there is great simplicity, almost a quaker-like rigidity of form, given to the machinery; no ornamentation, no rubbing away of corners, or polishing; but the precise, accurate and correct results. It was that which gratified me so much at Colonel Colt's, to see the spirit that pervaded the machines—they really had a very decided and peculiar character of judicious contrivance.[83]

That Nasmyth was struck by "correct principles," "common sense," "simplicity," "quaker-like form," "no ornamentation," and "the spirit that pervaded the machines" indicates a man interested in much more than the technical aspects of arms making. The operating of machines was inextricably linked with the culture and organization of work and the vitality of a production unit. Such a centrally organized production unit mass produced small arms with interchangeable parts, made maximum use of precision machine tools in sequential operation, was innovative, and facilitated the diffusion of new technologies. As we shall see later, this system provided a compelling model for the modernization of the Russian small arms industry. However, for the American system to reach Europe it was not enough for modern weapons to be designed and mass produced in factories organized according to "correct principles." They also had to be marketed.

International Marketing and the Transfer of Technology

To maintain a nation in arms and a nation of armorers requires funding by the national treasury. This burden has been great enough in the industrially advanced nations; in smaller, poorer, or industrially underdeveloped nations, the burden of military competition has been crippling. At the same time, as a study of arms production in developing countries argues, "For any nation, being in a position of having some other nation choke off needed arms supplies in order to get it to alter its behavior is unacceptable."[84] Because of this threat, most developing countries have followed the French model developed in the seventeenth century of royal (or government) monopoly, especially in the production of small arms.[85]

The arms industry has provided examples of the transfer and domestication of technology. To make the weapons alone requires the transfer of operational, technical, logistic, and managerial skills. The native repair and production facilities then become part of indigenous capacity and infrastructure. Studies of international technology transfer have delineated a variety of forms by which transmission of information or of products and processes embodying that information has been achieved in both military and civilian sectors. Such forms have ranged from simple export of a product or equipment to direct investment and the franchise of trademarks and production know-how. Foreign technicians who bring their expertise and natives who have studied modern technology at home or abroad provide the human element in the transfer of technology. In this way, an outmoded plant staffed with traditionalist workers and managers with little incentive to improve may begin either to produce goods that could not otherwise be produced or to produce existing goods more cheaply.[86]

The rapid domestication of complex arms manufacturing capability has always proved difficult. In more fortunate developing countries, the existence of a civilian industry that can be rapidly adapted to defense production facilitates the arms industry. The so-called dual-use technology can benefit the military sector.[87] Where highly negotiated environments have stalled the development of civilian industry, governments have created novel production organizations such as the *kuan-tu shang-pan* enterprises in China, designed to further the development of military and civilian technology by attracting the available private investment capital and managerial talent while retaining ultimate state control.[88]

Just as diffusion of innovation within an individual nation has met many obstacles, international transfer of technology has likewise faced many barriers. In non-Western countries with cheap labor and little capital (such as China), the absorption gap—the effective incorporation and assimilation of foreign technology—is often widened by the limited opportunities for the introduction of capital-intensive Western technology.[89] This applies particularly to the very costly mechanization of production. When foreign products are put into production, the resulting redesign to meet native standards causes delays in starting up production. Refuting a theory of substitution developed by such diverse thinkers as Leon Trotsky and Alexander Gerschenkron—which states that backward countries substitute advanced techniques for domestic deficiencies in capital formation, entrepreneurship, and labor force skills—a study of Soviet industry soberly points out that it is "often impracticable . . . to attempt to introduce into a society which is at a fairly early stage of industrialization those elements of technology and production organization that represent the most technically advanced features of industry in other countries. Modern industrial methods cannot be taken over full-blown from foreign models."[90]

American companies have been particularly imaginative and aggressive in their attempts to export technology. They not only have catered to but also have created foreign demand, especially for market-oriented facilities where a large existing or potential demand has existed for goods.[91] One of the earliest examples of aggressive sale of products and technology occurred in the mid-nineteenth-century small arms industry.

In its early years the United States had been the recipient of technology transfer. Designs had been commonly borrowed from French muskets. An American ordnance expert surveying European armies in the 1850s wrote that French muskets "are so well known as to render unnecessary a particular description of them except with regard to recent alterations and experiments."[92] Prior to 1866 the United States had imported arms in large quantities; exports were merely nominal. This situation changed almost overnight.

Rearmament of major European armies with rifled breechloaders in the 1860s and the 1870s coupled with cessation of government orders and plants producing below capacity in the United States after the Civil War provided the opportunity and the incentive for stepped-up exports of American arms. Between 1868 and 1878, the United States exported an average of $4.1 million worth of firearms annually. With laconic pride a contemporary American authority asserted, "With the single exception of the needle gun, every arm on a breechloading system used in Europe is of American origin, both in principle and application, a large proportion being of American manufacture."[93] A Russian observer was more loquacious, stating, "Just as European factories had earlier supplied America with arms, so America, in turn, is now the great industrial power. Its products are capable of glutting all the European arms markets and with little strain filling the enormous orders of European governments."[94]

More than any other arms maker, Samuel Colt publicized the American arms industry and advances in weaponry abroad. This publicity, as well as Colt's foreign marketing methods, is appropriately captured in two reliefs flanking a statue of Colt at Colt Memorial Park in Hartford, Connecticut. In "Parliamentary Testimony 1853," Colt lectures the British Parliamentary Select Committee on Small Arms on the virtues of his manufacturing methods. In "Royal Presentation, St. Petersburg, 1854," Colt, the first American arms maker to sell arms and arms-making machinery to the Russian government, demonstrates to the tsar and assembled dignitaries the virtues of his rifles and revolvers.

The legend surrounding Colt makes it difficult to assess clearly his contributions to the evolution of weaponry. This is as true of his foreign ventures as of his domestic exploits. Contemporary accounts of Colt's life, such as *Armsmear*, the biography commissioned soon after Colt's death by his widow, border on the hagiographic.[95] Folklore, such as that reflected in the aphorism "God created men, but Samuel Colt made them equal," abound. Colt was to a great degree responsible for the legend surrounding his own name, for he was not only a talented inventor and manufacturer but also a talented storyteller, promoter, and salesman. His exploits in these areas variously earned praise or scorn for his mechanical and manufacturing skills.

By 1850 Colt Patent Firearms Manufacturing Company was already the world's largest private arms company. On the eve of the Civil War the total capital of the company was valued at $1.25 million, almost three times greater than the nearest rival, the Sharps Rifle Company; at six hundred thousand dollars, the annual value of production was almost double that of Sharps, ten times that of Remington, and twenty-four times that of Winchester. In the decade between 1856 and 1865, Colt sold more than five hundred thousand revolvers and almost seven million rifles.[96]

Colt was one of the first Yankee arms makers to capitalize on foreign markets. Starting in 1849 Colt spent much time in Europe advertising his revolvers. He benefited greatly from the publicity generated by the revolver's exploits in America. According to one contemporary—in this case British—Colt promised to usher in the millennium:

> The name of Col. Colt is now patent to the whole world; for it is associated with one of the grandest schemes which ever entered into the mind of man, for the protection of human life where it is most exposed to danger. If the principle be admitted that the way to ensure peace is to be prepared for war, Col. Colt has devised a guarantee for perpetual harmony in the construction of a projectile which multiplies by six the powers of destruction.[97]

Although his aggressive promotion and sales techniques seem modern—and must have seemed especially daring to Europeans of the time—there was a decidedly traditional ring to many aspects of Colt's ventures in foreign lands.

First, Colt exploited the prizes and medals won by his weapons. Writing to Elisha Colt from Vienna in 1849, the gunmaker emphasized the need to exhibit his pistols wherever "they award gold medals. . . . These medals we must get and I must have them with me in Europe to help make up the reputation of my arms as soon as I begin to make a noise about them. . . . All these things go a great way in Europe."[98] In 1851 the Crystal Palace exhibition launched the first of many international exhibitions that were modeled on traditional trade fairs. The Crystal Palace offered to Colt, as later international exhibitions were to offer another generation of American arms makers, the opportunity to display his weapons and his manufacturing techniques.[99] Crowds gathered around the display to see Colt's revolvers made from interchangeable parts. According to one observer,

> The Old World was yet to be made familiar with their value; and the Great Exhibition of the World's Industry supplied the occasion. It is impossible to forget the sensation the array of Colt's revolvers produced. They were the one great feature of the American department of Crystal Palace. Demand, to an almost unheard of extent, followed upon the display. . . . The officer, and the traveller, and the emigrant, and even the careful housekeeper, alive to the intrepidity of the burglar, adopted this pistol.[100]

Second, Colt made a point of visiting world capitals and of obtaining audiences with political leaders. Once introduced to a head of state, or to a would-be head of state, Colt would present cases of custom-made, lavishly engraved weapons. In 1853, for example, he presented a cased pair of revolvers to Louis Kossuth as a "token of high regard and esteem," adding that he hoped Kossuth could one day visit his armory.[101] Although such tokens have the appearance of modern advertising, they were also given as gifts to patrons, one of the world's oldest forms of business. In return, Colt "came back to Hartford with his pockets bulging with orders from Russia, Turkey, and other powers."[102] At audiences with European royalty, Colt played the role of the "vigorous commoner of mechanical talent" rather well. No doubt such lavish gifts more than compensated for deportment unlike that of the European officer class. As Colt himself, perhaps disingenuously, admitted to Prince Murat concerning an audience with Napoleon III, "I must at the same time apologize to His Majesty and your Highness if my Republican habits have led me to adopt an unusual mode of addressing high Official Personages, as I am altogether unacquainted with the Etiquette of Courts."[103]

Whatever the nature of his "Republican habits," they did not prevent him from doing business with absolute monarchs. Indeed, Colt appreciated very well the importance of rank. Although Colt had never served in the American army, in 1851 Thomas Hart Seymour, the governor of Colt's home state, appointed him lieutenant colonel aide-de-camp in the Connecticut state militia. The title "lieutenant" was soon dropped, but the bogus "colonel" stuck until his death. The Belgian ordnance inspector Thomas Anquetil, critical of the exaggerated claims of the Colt revolver, was duly impressed with Colt's rank: "Before making his revolver, Colt was already fortunate; he was a colonel in the American service, a high rank in the American military." No doubt the military rank helped the common man to obtain "audiences with crowned heads of Europe."[104]

Third, and perhaps most germane to this study, Colt energetically promoted not only his weapons but also his manufacturing methods. After the Crystal Palace Exhibition, Colt sent presentation copies to Charles Manby, Secretary of the Institute of Civil Engineers. The gift was well placed, for at a meeting of the institute on 25 November 1851 Colt was invited to read a paper entitled "On the Application of Machinery to the Manufacture of Rotating Chambered Breech Firearms and the Peculiarities of Those Arms." In the paper Colt described the evolution of his weapons and the greater uniformity of parts, the reduced costs, and the use of unskilled labor that his machinery afforded.[105] Three years later Colt reiterated the significance of machine production and interchangeability to the Parliamentary Select Committee on Small Arms. Although a former employee disputed some of Colt's testimony (and Colt himself never claimed perfect interchangeability), three years later the new Enfield factory was furnished with American machinery.[106]

Fourth, although the prospects for sales in England were promising enough to appoint a London agent, Colt had no English patent and no means to protect himself from imitators. In 1852 Colt decided to set up a foreign branch plant in London with the help of Charles Manby. Fitted with American methods and machinery shipped from Hartford, the London armory was the first foreign branch plant of any American company.[107] The ever laudatory *Colburn's United Service Magazine* described the armory:

> The machines are small; but of the very best material, and being exclusively of American manufacture, enable us to reach a just estimate of the skill our Transatlantic brethren have attained in work of this important nature. . . .

We cannot, however refrain from indicating some of the specialties of the manufactory; and among the most prominent are, the beautiful order, cleanliness and quietude which prevail through the establishment. . . . The general result of such exactitude is an unapproachable perfection of manufacturing. Rectitude of intention is Colonel Colt's grand characteristic; and it is by enforcing a harmony and regularity of operations that his conscientious purposes are attained.[108]

On the continent, however, Colt licensed arms to be made by other manufacturers or, as we shall see later, by government armories.

Finally, Colt took advantage of the opportunity presented by the Crimean War to increase his sales. The new Enfield rifle was in short supply and the Parliamentary Select Committee, charged with ascertaining the best way to provide small arms, summoned Colt. "With the Crimean War in the brewing," a company history rather revealingly states, "Samuel Colt realized that the world was his market."[109]

Wherever there was a potential market for firearms, whether on the frontier of America or at the courts of Europe, Samuel Colt tirelessly promoted new weapons and new manufacturing techniques. According to one student of the American arms industry, "More than any other arms maker of his day [Colt] realized the importance of stimulating demand through aggressive sales promotion."[110] Other American inventors and arms makers followed a similar path in their attempts to publicize their designs and to get a share in foreign arms markets. The Crystal Palace Exhibition, and successive international exhibits, provided an entry into the British and European arms markets for many American gunmakers. Robbins and Lawrence and the Ames Manufacturing Company, in addition to Colt, won prizes at Crystal Palace for their displays and later received British government contracts. The accounts of subsequent exhibitions are filled with descriptions of new American weapons designs.[111] The opportunity to sell arms and machinery to Britain, and later to other European governments, was good business both for the government armory at Springfield and for the major private companies.

Foreign sales and government contracts were perhaps the best publicity for increased foreign sales and government contracts: one government's decision to adopt a weapons system became a model for prospective customers. Colt was not alone in capitalizing on this form of advertising. According to a contemporary survey of American manufacturing, "The best evidence of the high value placed upon the cartridges manufactured by Union Metallic Cartridge Company is the fact that the governments of Russia, Germany, France, and Spain have established manufactories for making this system of cartridges which has been adopted by them as ammunition in their respective countries."[112] Russia became a model for the adoption of American metallic cartridges by other armies. According to one Russian officer, Spain was using "Russian" cartridges made by the Union Metallic Cartridge Company and, like Russia before it, had ordered machines and skilled laborers from the United States.[113] And, of course, wartime orders provided the best publicity for future sales. Publicity was not always favorable, and American arms manufacturers quickly gained a reputation for their mercenary business practices. The London *Telegraph* reported,

American industry is profiting largely by the belligerent propensities lately developed in the East of Europe. Russia and Turkey are outbidding each other in the markets of the world for implements and agencies of slaughter, and it is in the United States alone that they find the means of gratifying their wishes promptly available. . . . On her side, Turkey was not far behind in the mad expenditure, having contracted with a Rhode Island company for 800,000 Martini-Henry rifles, of which more than half had been shipped. From other sources it was before known that the Porte was getting cartridges by the million from the same accommodating nation.[114]

As a result, advances in firearms and in manufacturing methods spread around the world. "For the next thirty years, these and other New England manufaturers supplied armories for practically all the governments of the world. Interchangeable manufacture spread everywhere."[115] Eventually it spread to Russia, although not without great difficulty. Due to defeat abroad and sweeping reforms at home, the military needs of Russia at mid-century were changing dramatically. The Russian government was receptive to new weapons and new methods to construct them. The following chapter will examine the state of Russian firearms on the eve of the Crimean War, the role of small arms in Russian tactical doctrine, and the government's efforts—particularly its business with Samuel Colt—to provide the army with rifles during the Crimean War.

SMALL ARMS IN PRE-REFORM RUSSIA
The Russian Colt

> "The study of the development of small arms in Russia shows that the arms of the Russian army lagged behind those of European armies, notwithstanding the talent of individual Russian gunsmiths, who left behind magnificent models in various museums and collections of old guns."
>
> —V. G. Fedorov, *Evoliutsiia strelkogo oruzhiia*[1]

● AN INCREASING ROLE IN EUROPEAN politics forced Russia to adopt western military technology. However, during the first half of the nineteenth century none of the innovations in weapons systems or in manufacturing discussed in the previous chapter took place in Russia. "The small arms used in the Russian army appear to be all made after models selected from those of other countries," wrote a visiting American ordnance expert in 1860.[2] Russia lagged behind Europe in economic and technological development; naturally this backwardness had an adverse effect on the Russian arms industry. However, as Walter Pintner convincingly argues,[3] during the eighteenth and first half of the nineteenth centuries, Russia's backwardness in weapons mattered little in the performance of the Russian army. More importantly, the weapons available conformed well not only to the other components of the military system but also to the greater needs of army and society. These needs were defined by

budgetary constraints, the nature of the peasant soldier, the weakly developed division of labor, the production, use, maintenance, and repair
of weapons, and Russian infantry tactics. Until disturbed by a series
of external and internal shocks at mid-century, the needs of the
army did not pose insurmountable problems to Russia's social and
economic structures.

Russian Small Arms: Systems and Soldiers

Throughout the first half of the nineteenth century, the standard
infantry arm was a smoothbore, muzzleloading flintlock musket, comparable to the British Brown Bess. The 1828 model, widely used even
at mid-century, used round balls and had a range of three hundred
yards but was not accurate beyond two hundred. Russian officers did
keep abreast of European developments in firearms, and, as the number of new models increased rapidly during the second quarter of the
nineteenth century, the army made extensive tests of new weapons.
In 1843 the Russian army began to convert the flintlocks to percussion
muskets according to Belgian and French models. Tests of rifles continued into the 1840s, and sharpshooter battalions adopted in 1843 a
model made in Liège and in 1851 the Ernroth rifle, designed by a captain in the Finnish sharpshooter battalion. The elongated Minié balls
were tested as soon as they appeared in Europe, but they were not
fit for the smoothbore muskets and consequently were not in widespread use during the Crimean War. Although the needle gun had been
adopted in neighboring Prussia, breechloaders were not highly regarded. (As we have already seen, Russian misgivings of breechloaders
were shared by many European armies.) According to one artillery officer, "In spite of the many clever attempts to construct a sturdy and
convenient breechloading mechanism, this system has proved to be
too complicated and accordingly is not sturdy or convenient enough
for a military weapon."[4]

Although it might seem at first glance that Russia was woefully
deficient in infantry arms, its choice of weapons systems made considerable sense given the related factors of budgetary constraints, the nature of the peasant soldier, and production capabilities. The latter factor will be analyzed in greater depth in the following chapter; suffice
it to say here that by the early eighteenth century Russia was largely
self-sufficient in firearms and, according to Pintner, "shortages of muskets were not a significant problem in the pre-reform era." This was
just as well, for by far the largest portion of the military budget was
devoted to subsistence expenses—officers' pay, food and clothing for

the soldiers, fodder for the horses—with a small, and remarkably constant, portion devoted to weapons.[5]

The small sums allocated for weapons put great pressure on what the editors of *Voennyi Sbornik* (Military review) called a chaotic procurement system, and shortcomings in this area may have been more detrimental than backwardness in weapons systems per se. Since weapons procurement came out of regimental budgets, and since these were niggardly and consumed largely by subsistence expenses, regimental suppliers understandably gave weapons procurement a low priority. Regiments tried to pay as little as possible for weapons, and the regimental suppliers regarded periodic trips to grimy government arsenals and to distant small arms factories as punishment. To make matters worse, regimental suppliers were frequently ill-informed about procurement rules and the qualities of firearms; for their part, armory distributors and inspectors tried to unload all weapons in stock, without regard for their serviceability:

> Naturally, in this state of affairs, the arms factories were interested only in supplying the number of arms ordered and not in insuring quality. The arms factories were convinced that the arsenal inspectors would not put pressure on them—that is, would not bite the hands that fed them—while the regimental inspectors would not be able to.[6]

No wonder that defects in muskets were cited again and again in the sources: screws not properly inserted, rivets placed badly, barrels fitted badly, stocks made with rotten wood, and lock parts mismatched.[7] No wonder also that Russian arms were not delivered in sufficient quantity or quality. As of 1 January 1853, only 532,835 of an authorized 1,014,959 infantry muskets were available.[8] Pintner's claim that shortages of muskets were not a significant problem in the prereform era would seem to be valid only in peacetime.

Clearly, given the constraints imposed by the budget and by the procurement system, the adoption of modern percussion muskets or rifled breechloaders would not have increased the availability of weapons in the field. The user of the infantry arm, the peasant soldier, imposed yet another constraint to modern military weapons. Recruited for a service of twenty-five years until the military reforms of 1874, the nineteenth-century peasant soldier, and his family left behind in the village, regarded military duty as an unmitigated disaster. Research into the attitudes of the peasant soldier by various scholars, and most imaginatively by John Bushnell,[9] suggests that training in the use and maintenance of sophisticated weapons was not easily achieved.

Many observers noted that Russian soldiers used and maintained weapons poorly. The military historian Fedorov claimed that "only with the adoption of the rifle was any attention given to the care and maintenance of expensive weapons."[10] There are many examples of shoddy care. For target practice soldiers used clay bullets that damaged the barrels. Screws and screwdrivers that fit improperly scratched the stock and barrel. Because the parade mentality dictated that arms were to be clean and shining, soldiers saw no need to dirty the gun with grease and oil. Frequent polishing thinned the barrels at the same time that frequent cleaning enlarged the bore; insufficient grease damaged the locks. To make matters worse, the soldier was supposed to pay for grease out of his own pocket. "Thus, weapons made without adherence to all the rules . . . and assembled less carefully than desirable were issued to soldiers, unfamiliar with the rules for maintenance, unable to afford grease, and lacking the proper tools. It is no wonder that under such conditions, weapons were frequently damaged and required frequent repair and shipment to repair shops."[11]

Work in regimental repair shops was little better than work done by soldiers. Before the Crimean War, according to *Voennyi Sbornik*, the gunsmiths listed on the regimental rolls learned their craft only in the regiment. Before military service these gunsmiths had been mechanics, wheelwrights, or carpenters, not locksmiths or stockmakers. Clearly, gunsmiths in the regiments lacked proper training and proper tools: "It is these dexterous mechanics and carpenters, having frequently built furniture, coaches, and sturdy pots and pans at home, these distinguished painters and wheelwrights, repairing firearms with the same chisels, hammers and saws with which they repaired wheels and shoed horses, who have brought our weapons to the state they were in recently."[12] Foreign gunsmiths, typically German journeymen rather than master gunsmiths, lured to Russia by the prospect of high wages were allegedly no better, despite the mystique associated with foreign skills. Although an 1855 regulation stipulated that work in regimental repair shops be guided by a civilian master gunsmith under the general supervision of a staff officer, repair, as well as fabrication and procurement, continued to pose problems.[13]

The difficulties in maintenance and repair were frequently cited by critics of new rifles and revolvers. An 1859 artillery textbook argued that the difficulty of keeping a revolver in repair counteracted its military advantages and advised that the new weapon be used very sparingly. Others noted that the slightest damage negated the most important quality of the rifle—accuracy. Even cavalry officers allegedly did not care for revolvers properly; and the sophisticated repair work could not always be performed by the regimental repair shops. Thus, the Small Arms Committee of the Ministry of War in the early 1860s advised arming junior cavalry officers with the less complex rifled pistols rather than with revolvers.[14]

Weapons, Tactics, and Training

Russia's human and material resources during the eighteenth and early nineteenth centuries shaped the country's tactical doctrine. Human resources were more important than material resources, and military art and "moral force" were more important than an abstract or scientific tactical system. The bayonet attack in closed formation and deep masses, protected by artillery, and the shock charge of cavalry were primary offensive tactics; the corresponding martial virtues to sustain the bayonet and cavalry attack were valor, blind obedience, and steadfastness; firepower, considered unreliable, was thought to sap the soldiers' courage. Such, in extremely general terms, was Russian tactical doctrine until the mid nineteenth century and beyond.[15] It resembled the tactical doctrines of continental military elites, and it also fit well the nature of the somewhat untrustworthy peasant soldier, the budgetary constraints and low priority given to weapons procurement, the backwardness in weapons systems, and the inattention to maintenance and repair of weapons.

More germane to our study, an underestimation of the effect of firepower continued to characterize Russian military thought in the pre-Crimean era. Cited frequently was Suvarov's aphorism: "The bullet is a fool, but the bayonet is a fine lad" (*"pulia dura, a shtik molodets"*). Baron N. V. Medem and Colonel F. F. Goremykin, professor of tactics at the St. Petersburg Military Academy, argued that musket fire was ineffective from distances greater than 150 yards, that small arms fire alone was insufficient to overcome a foe, and that only a bayonet charge could crush the enemy. Greater firepower, far from relegating "cold steel" to a secondary role, increased the importance of "moral force." General N. N. Muraviev believed that the breechloading musket, by intensifying the soldier's alleged tendency to fire off too many cartridges, "would do just the opposite of what was needed. . . . Troops having this weapon would cease to fight [hand-to-hand combat] and there never would be enough cartridges."[16]

Arguments such as these put forth in the Nicholaevan era were by no means off the mark. Firearms technology during the first half of the nineteenth century was rudimentary. Small arms were highly unreliable, especially in battlefield conditions. Russian smoothbore muskets had a range limited to three hundred yards. According to Fedorov, "Considering the short range and inaccuracy of smoothbores, it was natural that soldiers trained in Suvorov methods did not pay serious attention to long-range fire and regarded their guns as 'a machine for presentation.'"[17] However, even when improvements came rapidly in firearms technology, such arguments persisted even into the reform era.

Certain features of Russian infantry training and drill were a direct result of the available human and material resources and the principles of tactics applied. According to the military historian Fedorov,

> In the West, the soldier is trained to act on his own, while in Russia soldiers act in mass, do not take initiative, and are inflexible. When rifled arms were introduced in European armies, much attention was given to the soldier's flexibility, mobility, agility, action in open formation, and adaption to surroundings. At the same time the Russian army, raised on the scriptures of Suvorov which stressed bayonet force over accuracy of fire from long range, ignored the individual training of the soldier and the rapid adoption of rifled arms.[18]

Under Nicholas I, infantry drill had consisted largely of presentation of arms and of parades, the "Russian parade-mania," as John Sheldon Curtiss put it. Field training was virtually nonexistent, and soldiers had to practice musketry during rest time.[19]

The weakly developed division of labor in Russian society as a whole further limited the amount of time available for field and technical training. The army was obliged to undertake much unskilled labor that in societies with more division of labor might have been performed by the civilian sector. The regimental economy, so skillfully examined by John Bushnell, mirrored the society from which it came. According to *Voennyi Sbornik*, each regiment devoted excessive time to "economic" duties—agriculture, gathering wood and hay, construction—leaving little time for the soldier to master technical skills.[20]

One reason that the bullet was a "fool" in Russian tactical doctrine was that it invited the soldier to waste ammunition. As rate of fire became more rapid during the second half of the nineteenth century, this became an ever greater concern. The fear that the soldier would waste costly ammunition dictated training and supply in the pre-Crimean era, and economy and strict accountability in supply became the guiding principles. This further limited the already limited tactical advantages of fire. In the Nicholaevan era the powder and lead allocated for cartridges provided only ten ball cartridges and sixty blank cartridges per man. Artillery detachments that issued powder and lead were liable to the treasury for excess amounts. According to one American authority, "A strict accountability is kept with each soldier, to whom is given forty cartridges, for which he is personally responsible, unless in actual engagement with an enemy. Each night he must hand his empty shells to the armorer, who returns them to him in the morning reloaded, when they are inspected by his officer."[21]

Target practice left much to be desired. Clay bullets, which scratched the musket barrels, were used in place of lead bullets. In training and target practice, bullets were used more than once. (Allocation of cartridges for target practice became less stingy after the Crimean War, when instead of three cartridges, each soldier got 50 for smoothbores and 225 for rifles.) "The vast majority of infantry could not even shoot, other than volleys with blank cartridges fired during a ceremony. Ten live cartridges were distributed annually for firing, but even these were not spent on aimed fire." No wonder that it took a long time for Russian soldiers to train and that they were "especially slow at learning the use of firearms; even in crack corps, like the Guards and Grenadiers, musketry and artillery practice is, as far as accuracy of aim is concerned, very defective."[22]

The Failure at Crimea

Russia's defeat in the Crimean War sent shockwaves throughout a seemingly invincible system. Shortcomings in firearms contributed in important ways to Russia's defeat. To begin with, the deficiencies in training, maintenance, and repair increased the ineffectiveness of the arms that were available. Moreover, Russia's infantry arms were technically behind those of the Allies. Russia had achieved conversion of flintlocks to percussion muskets by 1852, but there were a sufficient number of muskets only for the army in active operations. According to a British observer, "The manufactories of Tula, Zlatoust, and St. Petersburg are hardly able to supply the loss of each day." Many soldiers were still armed with flintlocks, although these were in such disrepair that the Russian government was "on the point of seeing its armies with muskets that cannot fire."[23] The real weakness, however, was not in the lock system but in the barrel: while the Western powers had switched to rifled barrels with greater range and accuracy, the Russian infantry was armed almost exclusively with smoothbore muskets. General Totleben estimated that no more than 5 percent of the Russian troops in Crimea had rifled arms, while one-third of the French and one-half of the British infantry had rifles.[24] By the end of the war, British regiments were being supplied with the Enfield rifle firing the Minié bullet, the final advance in a muzzleloading arm. At the battle of Alma, the example most frequently cited, the 35,000-man Russian army had 2,000 rifles, while the French and British soldiers had 15,000 rifles. By the end of the war, 13 percent of the Russian infantry had rifles, 80 percent had percussion smoothbores, and 7 percent still had flintlocks. Charges the military historian Fedorov, "All these deficiencies were unthinkable among the Allies who were well armed and abundantly supplied."[25]

What this meant, of course, was that the Russian soldier was at a serious disadvantage in battle. Russian smoothbores had a range of only three hundred yards, and even then only one-fifth of the bullets hit their targets. The Allies, whose rifles had a range up to twelve hundred yards, could fire on the most distant artillery positions. Even the French smoothbores had a range of six hundred yards. Describing the disastrous battle of Chernaya Rechka, the American Minister to St. Petersburg, Thomas Hart Seymour, reported, "The Russians lost several of their best generals: they were picked off by the murderous Minié."[26]

The Crimean War demonstrated the growing importance of firepower. But Russian tactics, based on old weapons and static training procedures, did not respond well to the demands of the new weaponry possessed by the British and French. According to Cyril Falls, British superiority in musketry (not only in weapons but also in the training in their use) allowed infantry to operate in a thin line. The French used skirmishers backed up by columns. The Russians, in contrast, continued to favor the precision of the close-order movement and deep, heavy masses; the army paid little attention to marksmanship and skirmishing. "Their infantry was brave and enduring, but slow and thick-headed and precluded by its dense formation from making the most of its firepower."[27]

Artillery fared no better. Since British and French rifles could hit targets at long range, gun crews were no longer able to protect infantry. And since Russian units lacked the range to defend artillery, at Alma and Inkerman, according to Bruce Menning, "enemy riflemen had mercilessly shot down Russian gun crews who dared accompany infantry into attack."[28] Not surprisingly, projectiles caused few casualties in artillery; rifle fire was the cause of almost all casualties. As H. Liddell Hart concluded, "The Russians handled large formations in parade-ground style and moved about the battlefield in a densely massed way that took no account of improvements in firearms."[29]

After the outbreak of the Crimean War, the Russian government took several measures to supply a greater number of modern firearms. First, domestic technical advances were made during the course of the war. The calibration of the Russian smoothbore muskets was reduced. In 1854 the army began converting smoothbores into rifled bores using the Minié bullet, and in 1855 the cylindrical Neissler bullet was adopted. Although not without problems, this new system doubled the range of fire to six hundred yards and improved the firepower of the infantry.[30] The number of rifles distributed to the infantry was increased from twenty-six per regiment to twenty-six per battalion. To complement these technical changes, officers were given additional training in the use and care of rifles.[31] Although the range and accuracy of the Russian smoothbore still lagged behind that of Allied rifles, and although rapid fire was not achieved with the muzzleloaders, the limited advances caused *The Times* to predict, only one year after the conclusion of hostilities, that "should Russia be involved in another war her army will, in the use of small arms, not be found so inferior to the rest of Europe as it was in the past."[32]

Second, the government tried energetically to purchase arms abroad. This was, in part, due to the insufficient facilities at the three state-operated small arms factories and the lack of private small arms factories. Foreign purchases were not, in and of themselves, uncommon. The British also turned to foreign sources at this time and the United States was soon to do so during the Civil War. Large orders from the army had forced the Russian government to order abroad in the past, and it was customary for Russian small arms factories to assemble parts purchased abroad.[33] Wartime conditions, however, made it much more difficult to procure arms from traditional suppliers. In 1854 the Russian government ordered fifty thousand smoothbores and ten thousand rifles from the Liège firm of Fallis and Trapman, but the British, engaging in preemptive buying at Liège, ordered sixty thousand rifles. Liège gunmakers, taking advantage of Russia's acute need, raised the prices for Russian orders and shipped inferior arms. The high cost and inferior quality has remained a sore point among Russian military historians to this day.[34] To compound these problems, Prussia closed its borders for arms shipments to Russia in April 1855, at a time when the Baltic Sea was considered unsafe for shipping valuables. Only three thousand Belgian rifles reached Russia before the end of the war.[35]

It was natural that the Russian government turned to Liège with small arms orders at the outbreak of the war. As one of the oldest and most famous centers of small arms production, Liège had been sending firearms and gunmakers to Russia for a long time. However, for various reasons, Liège could not satisfy the Russian army's needs for small arms, and the government was forced to turn elsewhere. Just as the British Parliamentary Select Committee on Small Arms recommended orders from the United States, so Russia also turned to the relative latecomer in the manufacture of firearms, and in particular to America's best-known arms maker, Samuel Colt.

Arms Procurement in the United States:
Russia and Colt

Limited commercial contact between Russia and the United States during the first century or so after American independence took place in an atmosphere of friendly political relations. This atmosphere was fostered by a shared hostility to England, the young American government's avoidance of "foreign entanglements," and the indifference of American public opinion to Russia's role in Central Europe. Minister Thomas Hart Seymour summed up the official position of Washington:

> Whatever we may think of Russia in its character of a despotic power, hers is a government which has been uniformly friendly to that of the States and we can continue our trade with her and improve it if we please, without either endorsing her policy or compromising our republican principles.[36]

Republican principles did not get in the way of business, and an atmosphere of mutually useful cooperation existed because of Russia's interest in American technology and because of the perception of American statesmen and businessmen that Russians were also committed to progress and prosperity. Like American entrepreneurs, the Russian tsar was allegedly eager to harness unused resources.

> At the present moment the earnest desire of the Czar is to connect by telegraph and railroad all the important points of his vast dominions. . . . In working out her destiny, Russia will therefore assist America in realizing her own. Nicholas has already shown not only a willingness but an eagerness to import American enterprise into his empire and neither our interests nor those of the republican principles we uphold require us to draw back. We know that a wide field has been open to physicians, engineers, mechanics, and manufacturers from the United States; and who can foresee all the numerous benefits that will be derived from such a connexion, to a furtherance of liberalism in one country and the progress of commerce in both?[37]

Further down the social hierarchy, business promised a happy union between muzhik and mechanic:

> Who knows now what great and good influence in the cause of Freedom and Reform is exercised by the mingling of our mechanics with the peasantry of the Russian empire. Who knows but in a few years the now Russian serf, may stand a freeman at his own cottage door, and as he beholds the locomotive fleeing past, will take off his cap . . . and bless God that the Mechanics of Washington's land were permitted to scatter the seeds of social freedom in benighted Russia.[38]

Despite the favorable political climate, the eagerness of American entrepreneurs to sell products or innovations to Russians, and Russian curiosity about American technology, there were many obstacles to mutual trade. Neither country had a large foreign trade. Russia's foreign trade, for example, was approximately 30 percent that of France, 18 percent that of Great Britain, and only slightly larger than that of Austria, a country with one-third the population. High tariffs characterized the reign of Nicholas I; and the American trade increased only after the lowered tariffs of 1857.[39] The social, political, and economic order posed a formidable obstacle to Russian entrepreneurship, and incorporation presented obstacles to both Russians and foreigners. The Ministry of Finance and the State Council had to approve the corporate charters, the tsar had to add the seal of approval, and the charters had to be officially published before a corporation could begin operations. Although Russian patent law did not distinguish between Russian subjects and foreigners in the issuance of patents and in the granting of rights to establish manufactories, a foreign-owned manufactory was taxed higher unless the patent was transferred to a Russian subject, or unless the foreigner became a Russian subject temporarily. In addition, although the Ministry of Finance's procedure for receiving patent applications seemed simple, according to American officials in Russia, cases dragged on for a long time, to the detriment of American inventors. "American inventions are so highly esteemed, that with greater facilities for their introduction, a lucrative field would be opened to many meritorious ones which are yearly made public in the United States."[40]

The Crimean War made business relations more difficult and the shipment of goods more treacherous. However, the war also made Russia an attractive market for the makers of arms and ammunition. American consular officials in St. Petersburg were certainly not discouraging the arms trade:

> Contracts might be made for supplying machine made rifled muskets, which would be preferred to the Liège hand made. . . . *This government could be easily induced to order an extensive set of machinery* for making small arms, like that made for the English government.[41]

The *Times* claimed,

> The Russians receive great supplies of arms and ammunitions from [New England]. . . . If the *Samuel Appleton*—which vessel everyone here believes to have had arms and ammunition for her cargo in part—could get in [the Baltic Sea] ahead of the English, it is not unreasonably inferred that there will be enough time for operations at the close of the season to admit of

some dashing skipper landing powder enough at Port Baltic to blow all Europe to pieces.[42]

Indeed, the perceptive consular official was already one step behind the Russian government and Samuel Colt; America's largest gunmaker had just set sail for St. Petersburg to meet with Nicholas I, Tsar of All the Russias.

Russia's Initial Contact with Colt

Colt had come to the attention of the Russian military long before the Crimean War. From 1837 to 1840 Rusian naval engineers studied American methods of building and arming ships, supervised the construction of the steam frigate *Kamchatka* in New York, and followed Colt's experiments in coastal defenses and submarine explosives.[43] In 1841, the Russian Minister to Washington, Alexander Bodisco, who had learned of Colt's revolvers and their use in the Florida campaign, arranged for Captain Matthew C. Perry, then commandant of the New York Navy Yard, to escort a visiting Russian naval commission to Colt's Paterson factory. Captain Ivan I. Von Schantz, who was soon to take command of the *Kamchatka*, reportedly proposed that Colt "place his inventive talents at the service of Nicholas I." Colt took advantage of the opportunity to present the Russian naval commission with one of his revolvers and thereby to spread his fame.[44]

During the early 1850s the Russians used a variety of channels through which to familiarize themselves with American arms in general and with Colt's products in particular. In addition to the initial contact established by the naval commission, the Russians followed the designs and tests of new weapons systems through the foreign military press, established contact through foreign commercial agents in St. Petersburg, dispatched military agents to Europe and the United States, and pursued contacts through diplomatic channels. Although there is no evidence to suggest that the effort was highly coordinated, it paved the way for more systematic arms procurement and system adoption after the Crimean War.

By 1850, Russian officers, like their European counterparts, had read accounts of the Colt revolver. Captain Alexander Gorlov, who played an important role in Russian procurement and design of small arms, devoted much attention to Colt in an article on revolvers in *Artilleriiskii Zhurnal* (Artillery journal) in 1855. Gorlov credited Colt for being the inventor of the first practical revolver. This recognition can be found in other articles by Russian officers of the period, and this fact alone added to the revolver's esteem. Gorlov narrated the history of

Colt's revolver and provided his readers with a translation of the technical description in Colt's 1849 patent for improvements. After summarizing the results of tests and material available since the 1849 patent, Gorlov compared the Colt models with their "numerous imitations," particularly with the English Adams model. Although the latter was acknowledged to have a greater rapidity of fire, Gorlov gave the overall advantage to Colt primarily because it could achieve greater accuracy.[45]

In another description of revolver systems in 1855, K. Kostenkov stated that while

> the honor of invention of the revolver cannot be ascribed to Colt, he is responsible for its improvements. The reliability of the mechanism, its durability, the smooth and quiet cocking, the accuracy of sighted fire, and the firepower of the bullet are the major merits, compared to which the deficiencies—the weight and the inconvenience in holding the revolver—are inconsequential. And even these deficiencies are less serious than in the models of his predecessors.[46]

A notice in *Morskoi Sbornik* (Navy review) assured readers, presumably first-time revolver users, that the Colt model also had superior safety features. Another officer noted that Colt's revolvers, whose fame was energetically promoted by Colt himself, were "everywhere on sale and are now in use wherever there is military action."[47]

The Russian government took advantage of an existing link with a foreign manufacturing firm to establish personal contact with Colt. Ironically, given the diplomatic and military rivalry, it was an English firm that provided the first personal contact (excluding the Russian naval mission) between Colt and the Russian military. Firth and Sons, the famous Sheffield steel producers, were selling high-grade steel for barrels to both Colt and the Russian government. According to Firth's St. Petersburg agent, James Fretwell, visiting Colt's London factory in 1853, "the Russians, particularly the Emperor, are fond of anything new in firearms." Even more to the point for Colt, the Russians "have plenty of money and are favorable to Americans." Fretwell offered his services in the case that Colt were to extend his patent to Russia.[48]

At the outbreak of the Crimean War, Russian military agents were dispatched to Europe to order rifles and powder. As has already been stated, Liège had long supplied arms to Russia. General G. A. Glinka, a proponent of Liège arms and a critic of the allegedly dubious quality of American arms, was sent to Liège to procure rifles.[49] Colt, in characteristic fashion, was ready to take advantage of an opportunity. Colt's

Belgian agent, J. Sainthill, a patent attorney, presented to the Russian minister in Brussels drawings of Colt's revolver, a copy of an ordnance report, and an article by Colt himself. In return, the minister introduced Sainthill to General Glinka. Two weeks later Sainthill wrote to Colt: "I saw the Russian Glinka and had a long interview with him and removed several misapprehensions that he entertained about Col. Colt's arms." At the same time Colt met Edward Stoeckl, Russian chargé d'affaires in Washington, who had already received from Sainthill a patent application for Colt's revolver. Stoeckl not only expressed his desire to help Colt receive orders for arms from his government but also arranged Colt's first visit to Russia.[50]

Wartime Arms Purchases in America

Attracted to Colt by the latter's publicity and by his personal contacts, the Russian government wanted from the American gunmaker rifles, revolvers, a license to manufacture Colt models in Russia, and machinery. The seriousness of the Russian government is attested by the fact that it invited the Yankee arms maker to Russia three times; during each visit Colt had an audience with the tsar, grand dukes, and officers. During Colt's first visit in October 1854, Nicholas I, although at the time residing at the royal palace of Gatchina outside St. Petersburg, made a trip to the Winter Palace in order to meet Colt. Invited a second time in 1856, Colt attended the coronation of Alexander II. During his final visit in 1858 he made his most extensive and lavish revolver presentations—fourteen engraved and gold-inlaid 1851 and 1855 models with the imperial monogram—to Alexander II and the Grand Dukes Michael and Constantine.[51] Colt, in turn, wanted sales, publicity, patents, and revolver adoption. Although neither Colt nor the Russians got everything they wanted, the Russian government began an important business and industrial relationship that was to pay off in the coming decades.

Forced by the Crimean War to seek new supplies of arms, the Russian government dispatched Captain Otto Lilienfel'd, a small arms expert, on a secret mission to the United States to procure muskets. Learning that many American arms factories were operating at capacity in order to fill English orders, Lilienfel'd determined that the Russian government could contract only with Colt, who at the time was converting smoothbores into rifles, and whose London and Hartford combined plant capacity far exceeded that of any other private armsmaker. On 12 July 1855, Lilienfel'd and Colt concluded a contract for the delivery to Russia of 50,000 converted smoothbores by 28 April 1856.[52] The British were keeping a close watch, and one captain of the Royal Artillery reported to the Ordnance Office, "I understand that 25,000 of the old muskets bought up by Kossuth in the States, but which he was obliged to leave behind unpaid for, have been rifled by Colt for Russia and are now awaiting shipment."[53]

The Russian government was also interested in revolvers, Colt's initial claim to fame everywhere. Before Colt left St. Petersburg in 1854, the Russian Navy agreed to purchase 3,000 1851 model revolvers. Unfortunately for both the Russians and Colt, in April 1855, Prussia closed its borders to arms shipments to Russia.[54] This did not deter the resourceful Colt, and on 29 May his Hartford secretary advised Stoeckl that the order had been shipped to Antwerp. Neither Colt nor the Russian government heard of the arms again until August, when the ever watchful *Times* reported that a consignment of 145 bales of cotton en route from Antwerp to St. Petersburg had been confiscated at Aix-la-Chapelle on the Prussian border after it was discovered that each bale contained 24 revolvers.[55]

The confiscation of arms destined for Russia had an adverse effect on the musket contract the Russians had signed with Colt in July. At Lilienfel'd's insistence, the contract had stipulated full payment on delivery in Russia. Claiming that the confiscation of the pistols at Aix-la-Chapelle had forced him to take more precautions, Colt asked for an extension of the contract.[56] Lilienfel'd refused, stating that the purpose of converting the smoothbored muskets was to get rifles to the troops as soon as possible. Two months after the contract deadline passed, the Russians informed Colt that the contract had been breached and the order would be canceled. Colt claimed a compensation from the Russian government for his loss, arguing that Lilienfel'd had delayed inspection of the arms. According to Fedorov, delays in the inspection did take place but were due to the poor quality of the rifling, a plausible explanation in light of the assessment of one of Colt's own agents, Charles Cesar: "the [Russian] order was the very worst lot of arms . . . made at the Hartford Armory" due to "gross neglect in the fitting and inspecting departments."[57]

On 18 March 1856, the Peace of Paris was signed, ending the Crimean War. Although Russia no longer urgently needed converted muskets, Colt energetically pursued Lilienfel'd in an attempt to renegotiate the contract or to persuade the Russian government to take the arms "without alterations," a phrase that could be interpreted by the Russians to mean accepting arms despite their condition or accepting smoothbored muskets without rifling. Colt hoped that Lilienfel'd, then stationed in New York, would give a favorable report to his government on the condition of the muskets. "His report may be of service to me . . . in getting his government to purchase the muskets in their present state . . . and pay for them in New York."[58]

Colt might have been an annoyance to Lilienfel'd had not the Russian government been planning more than mere purchases of arms. The difficulties in procuring arms in wartime prompted the Russians in 1854 to purchase from Colt something more valuable in the long run—a license to make, or to assemble, Colt revolvers at the Tula and Izhevsk small arms factories. According to a report of the Committee on Improvements in Small Arms, by 1855 the Tula small arms factory had begun production of three models of the Colt revolver, and the first mention of officers being armed with revolvers appeared in an Order of the Ministry of War on 14 September 1855.[59] In his 1855 report Gorlov acknowledged that Colt revolvers were already being made at Russian small arms factories: "In those countries where Colt has not taken a patent, revolvers of his system are made at various small arms factories according to the specifications of American drawings and models. . . . So, too, three models of Colt revolvers are being made at our armories."[60]

A licensing agreement suggests one more purchase of interest to the Russian government: machinery. According to the Soviet historian of the Tula armory, who had access to unpublished material in Soviet archives, as early as 1855 Colt offered to provide the machinery and tools necessary to make eighty thousand rifles annually.[61] Somewhat overwhelmed and not convinced that this could be accomplished in wartime, the Russians were cautious. But in June 1855 Lilienfel'd began negotiations with Colt for a few machines to be installed on an experimental basis. An official United States military commission, visiting Russia in 1855, confirmed that "Russian officers are now in the United States to procure machinery for making rifle-muskets of the same calibre as those now in use, to carry the grooved Belgian [Minié] ball."[62] According to a contract signed 1 October 1856, just before he left Russia, Colt agreed to install thirty-four stockmaking and rifling machines at the Sestroretsk armory.[63] Colt hoped to get further orders for tools and machinery while in Russia in 1856:

> I have endeavored to impress upon the minds of these people here the importance to them of a full set of machinery for making rifles, if they have any at all, and they cannot start a good manufactory for less here than it has cost in England, now reported officially to be £150,000, and it will cost them about £50,000 more to get ready for making rifles.

Colt later boasted that the rifling machinery "furnished by us to the Imperial Government of Russia [is] of the finest quality of material, the most perfect mechanism and highest finish of any now in use."[64]

Colt's brother-in-law was sent to Tula to supervise the machinery's installation. Although a letter in the *Times* nervously claimed that Colt had visited Tula and that "the Russian government is to reorganize the Tula armoury and, as I heard, put an American as its head," this never happened.[65] A reorganization of the armory and technological transfer on a large scale were still in the future. Nevertheless, Russia's machinery purchases, like its arms purchases, marked an important step in the modernization of Russian firearms. At the same time, that both Russia and Colt received less immediate benefit out of the wartime procurement suggests important factors limiting arms modernization in Russia.

Russia and Colt: The Limitations of Wartime Arms Procurement

Russia's isolation during the Crimean War made arms procurement from traditional suppliers, such as Belgium, more difficult and expensive. Although American arms makers, especially Colt, had excess plant capacity and were enthusiastic about selling arms (to all combatants, as it turned out), acquiring new sources for arms was equally difficult and expensive. As noted, Colt's revolver shipment was seized at the Prussian border, and the production delays in the musket contract rendered the weapons useless in wartime. Although the Russian government showed a willingness to tap new sources of arms, no accounts of the weapons used indicate that new arms made much difference on the battlefield.

Given that Colt's rifles and revolvers received much praise in the Russian military press, it is still curious that the government, particularly in wartime, should not have ordered more of them. Expense and transportation risks deterred the government from ordering large numbers. However, four additional reasons dissuaded the Russian government from large orders: reservations about the effectiveness of new weapons, technical criticisms of Colt's innovations, foreign competition, and a budding native industry.

In 1861 the Small Arms Committee deemed that revolvers were too expensive and too complicated to be supplied to the lower ranks. In addition, the revolver mechanism, which required great care, was easily damaged and not easily repaired. Although the committee acknowledged the desirability of arming cavalry officers with revolvers as soon as further tests could establish the best system, for the time being cavalry officers were armed with a Belgian-designed breechloading rifled pistol.[66]

Although the inherent cautiousness and conservatism of Russian officials provide an easy target for the historian, new weapons did receive considerable study and testing. In 1856 the Committee on Improvements in Muskets and Guns considered models of carbines and revolvers from several gunmakers, including Colt. Colt had hoped to patent his new sight in Russia, but the committee reported negatively. Colt's six-chambered carbine was also considered deficient: in addition to being a muzzleloader, it was too heavy and sealed gas inadequately at the breech.[67] Five years later the Small Arms Committee considered several revolvers for adoption. The single-action Colt was judged to be too complicated in construction and too difficult to load and unload. The Small Arms Committee concluded, "The Colt has passed its prime. After the appearance of the other systems, it appears more complicated and less convenient in comparison."[68]

Such criticism of Colt's innovations adversely affected their adoption by the military. There is evidence, albeit scanty, that Colt revolvers found limited civilian use. It would appear from circulars and orders that by the early 1860s prison convoy guards were supplied with Colt revolvers, as were officers of the Third Section, the political police. By the mid-1860s, several St. Petersburg arms dealers were keeping a supply of Colt revolvers for sale to officers. A 1858 report by F. S. Claxton, the U.S. consul, suggests an additional potential use for Colt revolvers: "The fears consequent upon the agitation of the 'Emancipation question' having induced travellers and proprietors to provide themselves with repeating firearms, there is not at the present time five of Colt's revolvers on hand in Moscow—any manufacturer in the U.S. would find a ready sale for over 1,000 pistols of this description."[69] Documentary evidence, however, indicates that revolvers were seldom used by private individuals.

Being backward in the development of new weapons systems such as revolvers, the Russians were in a position to compare many foreign models; this foreign competition, of course, did not benefit Colt. Although Gorlov and Kostenkov relied on preliminary tests of the Colt-model revolvers made in Russia, such material was scanty in 1855 and to a great extent they based their articles on Colt's own test results. One year later, Gorlov revised his estimation of the Adams revolver, because he learned through *Colburn's United Service Magazine* that Adams had improved the revolver to make it more convenient to use; Colt, he added, was still working on improvements.[70] The Small Arms Committee, and Russian small arms experts in general, found that the double-action Adams-Deane, offering more rapid fire, and the French Lefaucheux "had many advantages" over the single-action Colt.[71]

The Russian government, of course, was not completely dependent on foreign arms suppliers, even during wartime. Native arms designers met the challenge of new weapons and offered their own improvements. Although the Russian tsars received Colt and highly praised his revolvers, in setting-up production of large quantities, preference was given to a Tula model revolver equipped with a raised sight that permitted greater accuracy.[72]

In its caution regarding foreign weapons and in its adoption of new rifle and revolver systems, the Russian government was driven by three overriding considerations. First, budgetary constraints severely limited the expenditures in new weapons. While this was not a serious problem in peacetime, it did mean that urgent wartime needs were fulfilled slowly or not at all. Second, the nature and training of the peasant soldier demanded that the weapons system adopted be simple to use and have a long service life. This, too, made sense in peacetime and at times when technical changes in weapons proceeded at a slow pace. However, neither was the case in mid-century. Consequently, to an American observer of Colt's business at the Russian court, it seemed that "to all representations as to the importance of improved arms the answer was, 'Our soldiers are too ignorant to use anything but old Brown Bess.' The result was that Russian soldiers were sacrificed by the thousands; their inferiority in arms being one main cause of final defeat."[73]

Ease of domestic manufacture was a final criterion in system adoption. Criticism of Colt's musket innovations was not merely technical. In the words of the Small Arms Committee, "At first appearance the American sight is indeed simple and strong. However, the difficulties in manufacturing constitute a significant deficiency."[74] The difficulties in supplying Russia's wartime needs suggest that Russia's native firearms industry was unable to meet the challenge posed in mid-century.

THE RUSSIAN SMALL ARMS INDUSTRY

● FOREIGN GUNMAKERS, ESPECIALLY THOSE AT Liège were an important source of firearms for the Russian military. Because Russia had a native firearms industry, although it borrowed system design and innovation, the state was not dependent on foreign sources for its military needs. Even though private producers of small arms existed, the Russian firearms industry essentially consisted of government arms production. The head of a British military delegation in 1867 put it succinctly: "The almost total absence of private manufactories, capable of giving assistance in time of emergency, throws a double responsibility upon the government."[1] The purpose of this chapter will be to trace the development of the government armories, of the labor force in the firearms industry, and of the progress of technology through the middle of the nineteenth century. An analysis of such aspects as legal status, corporate organization, recruiting procedures, shop practices, wages, and working conditions will illuminate several features of government-labor relations, of prereform Russian society, and, most important, of the production of weapons. As will be shown, the gunmakers themselves played an important role in the organization of work and the application of new technologies in the state armories.

Because of the larger number of gunmakers and because of its longer tradition, the Tula armory commands the greatest attention. The statue of Peter the Great by R. R. Bakh dominates Tula's central square. The hammer has already struck the anvil, the work has been started, and the artisan-tsar, sleeves rolled up, dressed in the iron forger's apron, is pausing for a rest. The Tula craftsmen who became great manufacturers—Batashev, Moslov, and especially Demidov—are, like Peter, larger than life. In 1812 the patriotic Tula gunmakers heeded the call of the Russian crown and delivered arms to a beleaguered nation.[2] According to a nineteenth-century account of the armory, "There are few individuals travelling in the area, including even the ladies, who are not eager to visit [the armory], despite the risk of sullying their clothing."[3] It is to the history and administration of the government armories that we now turn.

The Government Armories: History and Administration

Settlements of metalsmiths, such as the area in Moscow known as Bronnaia sloboda, emerged in the fifteenth and sixteenth centuries. Metalworkers in these settlements made guns for private customers and at the same time did contract work for the tsar, a feature of the Russian small arms industry that continued into the nineteenth century. In 1595 Boris Godunov established the first state gun foundry near iron ore deposits south of Moscow in the town of Tula. Over one hundred years later, on 15 February 1712, Peter I decreed that "for improving the method of arms production a site should be found at the gunsmiths' village and factories thereon erected."[4] Thus the first state small arms factory at Tula commenced the fabrication of service arms. In 1724 a second state small arms factory opened at Sestroretsk on the Gulf of Finland, near St. Petersburg. A third small arms factory commenced production in 1807, when the government reopened an iron foundry, established in 1760 but nearly destroyed during the Pugachev rebellion in 1774, on the Izh river in the Ural Mountains. According to Russian law, small arms factories fell into the category of industrial establishments operating exclusively under the auspices of the state.[5]

Although the small arms industry was essentially a government operation from the beginning, this was by no means the only possible strategy of industrial development. Private entrepreneurs did exist. According to an authority of seventeenth-century Russia, when the government contracted with a merchant to provide a commodity, the merchant was given an advance on future delivery. This advance provided the capital necessary for the undertaking, and the government "acted as a commercial and industrial development bank."[6] Even before Sergei Witte became Russia's Minister of Finance and best-known industrial developer, such government pump-priming of private entrepreneurship had a long tradition.

In theory the Russian government could have provided capital advances to firearms manufacturers, as did the American government, thereby seeding a private firearms industry "capable of giving assistance in time of emergency." Instead, in the eighteenth century the government created an alternative to private industry for the manufacture of many products. The government established the so-called possessional factory, a hybrid between privately owned and state-owned enterprises. Possessional factories arose as a compromise between the need for forced labor to encourage Russian industry and the gentry's monopoly of serf ownership. An entrepreneur, usually a merchant, owned the factory, but the government owned the serf labor force, which was bound to the factory. The government regulated relations between the factory owners and the serf labor force. The owner did not have the right to break up the factory, to sell the so-called possessional serfs separately from the factory, to transfer the serfs to another factory, or to change the nature of production. Any augmentation or improvement of plant and facilities required approval from the government. Possessional enterprises were obliged to deliver a certain production quota to the government at fixed prices; any surplus was disposable at the owner's price. The factory owner could curtail production but until 1835 was obliged to provide laborers with work and wages, even when market forces might have dictated laying them off.[7] Clearly, this was government intervention in the extreme. Although possessional factories became less necessary toward the end of the eighteenth century and declined dramatically by the early nineteenth century, they provided an alternative model to private entrepreneurship for the fostering of new industries.

In theory, the government could have created "possessional small arms factories," but, to stimulate a native small arms industry, the government neither provided capital advances to private entrepreneurs nor created possessional factories. Instead, from the time of Peter I the government wholly financed, owned, and directed this important branch of industry. As we shall see, control over resources, labor, procurement, wages and prices, and innovation was concentrated in the hands of the central authorities in St. Petersburg. This created a considerable imbalance between the center and the peripheral local armory authorities. Since alternatives to central control did exist, it is clear that the needs of national security coupled with the government's mistrust of its subjects to deliver weapons in wartime dictated state ownership and control of the small arms industry.

Until the creation of the ministerial system under Alexander I, the three small arms factories were run by different branches of the government. Sestroretsk, for example, was originally intended to provide firearms for the navy and at first was run by the Admiralty College. Izhevsk was a part of the Department of Mines. Only in 1808 were all three small arms factories transferred to the newly created Ministry of War.[8] Despite their location in different government departments during their early years, the government armories had a common internal organization. The oldest and best-known armory, Tula, provides a good example of the common internal organization.[9]

The quartermaster general was in charge of the Ordnance Department, which was responsible for the administration of the armory. The commander of the armory acted as director and commanding officer, responsible for the factory's entire operation, including the production, labor force, and finances. The commander chaired the governing board, the highest administrative body of the armory. While the governing board could act in many instances without the approval of the commander—contracts could be concluded, for example—the commander could make only a few minor decisions without the approval of the governing board. As a consequence, the commander alone was only rarely able to make significant changes in the operations of the armory. The commander had two assistants for business and technical matters. The latter assistant was in effect the chief engineer, directly responsible for the quantity and quality of work and also for the various gunmaking shops. Subordinate to the chief engineer were the mechanic, the architect, the chief superintendent, the chief inspector, and officials of the various shops. Most important of these officials were the foreman and the shop stewards. The actual supervisor of the shop was the foreman appointed by the factory administration. Usually officers, the foremen were part of the factory administration proper, while the shop stewards were part of the shop organization of the gunmakers.

One of the principal rationales for a state-run arms industry is to facilitate, even to control, arms procurement. A government that cannot easily procure arms or control the arms procurement process in wartime at best becomes dependent on rapacious producers and middlemen, and at worst becomes defenseless. Did the Russian government devise a reliable system to acquire the means to wage war successfully?

Arms procurement was a cumbersome procedure involving a high degree of central planning, disorganized bureaucracies with overlapping responsibilities, and inefficient and nearly autarkic producers. Given Russia's poor transportation network and the lack of specialization of production, most factories, including the government armories, were nearly self-sufficient. Within the Ministry of War, the department responsible for the weapons purchasing, procuring, and distribution was the Chief Artillery Administration. Sharing responsibility for these activities was the Quartermaster General of Ordnance, superior to the Director of the Chief Artillery Administration. Within the Chief

Artillery Administration, the artillery and small arms committees were responsible for testing all new weapons systems, and the Inspector of Armories was responsible for determining plant capability.[10]

Each year the Military Council of the Ministry of War appropriated a certain amount of money for small arms to the Chief Artillery Administration. At the same time, each armory provided information to the Inspector of Armories regarding the number and types of weapons the armory was capable of producing. Based on information presented by the Inspector of Armories as well as on actual weapons needs, the Chief Artillery Administration every September submitted a report containing the annual order for each armory to the military council for final approval. Each armory, in turn, submitted orders for materials needed from the artillery administration as well as from other departments, such as the Department of Mines. This, then, was the annual plan (the word used at the time) for procurement, which identified the product, the quantity, the price, and the methods of delivery.[11]

Receiving the annual order from the artillery administration, the governing board of each armory provided the information to the chief engineer, who divided the work by shop according to the number of worker-weeks per year. This calculation yielded the number of finished weapons each shop needed to produce per week. After the chief engineer added to this figure an estimate of the anticipated number of defective weapons, usually very high, each shop received its final weekly quota. This weekly quota was then divided among the gunmakers as equally as possible.[12] Thus, although orders were initiated from above, the armories and even individual shops put considerable constraints in the process itself. Reports on armory capabilities dictated the final orders of the artillery administration, and the sources contain no examples of rejection by the artillery administration of armory capability reports.

The amount of weapons in combat readiness on the eve of the Crimean War suggests that annual targets were not always met. Indeed, from 1825 to 1850, according to the Soviet historian N. N. Kononova, the three armories together produced an annual average of 65,000 small arms, approximately one-half of the average annual state order of between 105,000 and 135,000 guns. Occasionally an armory would produce only a few. In 1847, for example, Sestroretsk made only 190 weapons in a plant capable of producing 40,000 weapons.[13] Although defective weapons is the reason most frequently cited in the sources for shortfalls in deliveries, production problems went deeper. For example, despite government control of the three small arms factories, standardization among them did not exist. Tula parts did not match those of Izhevsk, and so on. Accusations of poor management occur repeatedly in contemporary studies. Although armory management will be an issue addressed in the final chapter, here the observation of a British visitor to Tula in 1856 will suffice: "the weapon factories worked there day and night during the war, and as no proper control was exercised upon their products, they delivered quite worthless goods . . . thanks to the rascality of the men who preside over the manufacture of weapons at Tula."[14] The procurement system, although seemingly centralized and well controlled, could not compensate for production problems associated with the available labor force and technology and with the organization of work.

The Labor Force

In the eighteenth century, when industrial skills were in short supply, the Russian government chose not to rely on the open market to provide the labor force to develop native industries. The state devised an ingenious system to meet its labor needs: a peasant labor force indentured to possessional factories. In a similar manner the government indentured a labor force to the state armories. This labor force can be divided into two distinct groups—the regular laborers and the gunmakers themselves.

Like the possessional serfs, the regular laborers were assigned to the small arms factories. At all armories, the laborers were a varied group consisting of blacksmiths, carpenters, mechanics, stonemasons, painters, and sawyers. Unskilled laborers gathered wood, stoked fires, dug coal, prepared iron and wooden blocks for barrels and stocks, and hauled supplies. In many respects the legal status and obligations of these unskilled laborers mirrored those of the state peasants. Flight from the factory was a felony, and for this crime laborers were tried in military courts. Like state peasants, the laborers at Tula paid a quit-rent (*obrok*) to the factory; Sestroretsk laborers performed a certain quota of labor. Unlike state peasants, the regular workers permanently assigned to the factories were exempt from conscription. At Izhevsk the laborers were mostly non-Russian—Udmurty, Tatars, Mariitsy ("Cheremisy"); many of the Russian laborers were exiles. They lived near the factory and owned small plots of land. In addition, they were permitted to engage in outside trading activities; the most prosperous individuals at Izhevsk controlled the local grain and salt trade.[15]

The Gunmakers' Corporate Organization and Legal Status

The state had an understandable interest in gunmakers because of what they could produce. In a country where craft skills, especially in metalworking, were in short supply, gunmakers' skills were a scarce resource. One might assume that the practitioners of a highly skilled and scarce trade could command considerable occupational autonomy and the privilege of dictating terms of trade. The state, after all, needed to maximize the use of this human resource and to ensure that the best gunmakers would practice the trade. However, rather than rely on market forces to provide these scarce skills, the state, as it did in many other areas of national life, preferred to indenture gunmakers to the factory. Therefore, gunmakers were no more autonomous than were other workers assigned to the armories.

The gunsmiths at the three government small arms factories were organized into guilds or shops (*tsekhi*) and work teams (*arteli*). The shops and work teams were an important feature of Russian labor. The work teams were associations of equals that formed spontaneously to fulfill a specific task. Labor resources and responsibility were collectively shared, and under serfdom the work teams were self-sufficient. Shops and work teams bore a superficial resemblance to western corporate structures, although the Russian labor association emphasized mutual support and the sharing of sacrifice.[16]

Shops in the gun trade were named according to the gun component produced, and separate shops also existed for making tools, equipment, and side arms (armes blanches). Each shop consisted of twenty work teams. Work teams, made up of one or two masters, one or two journeymen, and a few workmen, were organized according to trade. In addition, each work team selected a boss to supervise the work and an inspector. Each shop selected its own shop steward to be responsible for allocating supplies and raw materials, for collecting finished parts from the various work teams, and for distributing wages. As already indicated, the actual supervisor of the shop was the foreman appointed by the factory administration. Work team bosses and shop stewards, although paid by the society of gunmakers, were administrators, not gunmakers. Finally, the civil and criminal corporate affairs of the gunmakers were managed by a guild council, analagous to the city magistrate, and its elected representatives.[17]

In return for making arms for the state, in the eighteenth and early nineteenth centuries, the government awarded certain privileges and immunities to the gunmakers not granted to the state peasants or possessional serfs. Holding land and exemption from the soul tax and conscription were undoubtedly the most important. The Izhevsk and Sestroretsk armorers were included among those in military service; however, unlike laborers in the work teams, the gunmakers did not have to pay quitrent. The Tula gunmakers had a higher status. According to the 1782 charter of the Tula factory, Tula gunmakers

Alexander II, Emperor of Russia, 1855–1881. His coronation in 1856 was attended by Samuel Colt. The beginning of his reign is usually called the era of the Great Reforms.

Count Dmitrii Miliutin, Minister of War, 1861–1881. During Miliutin's tenure in office, Russia adopted universal military service and the reserve system. It was also during this time that Russia modernized its weapons and its government armories.

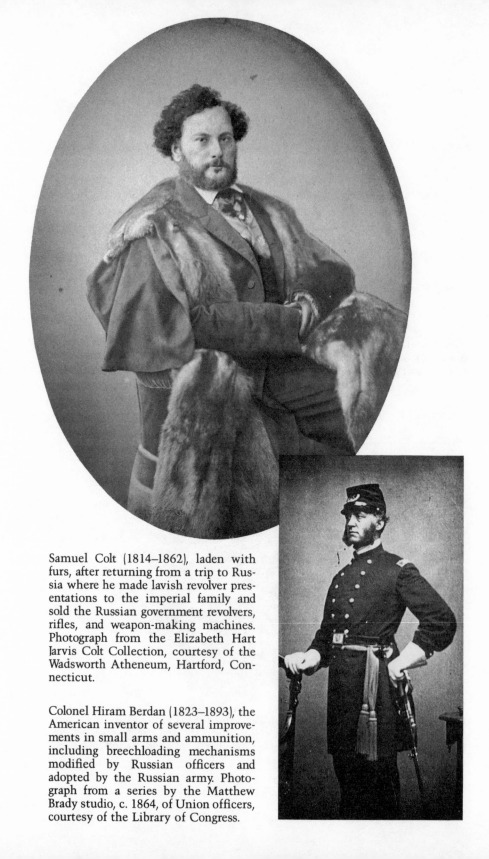

Samuel Colt (1814–1862), laden with furs, after returning from a trip to Russia where he made lavish revolver presentations to the imperial family and sold the Russian government revolvers, rifles, and weapon-making machines. Photograph from the Elizabeth Hart Jarvis Colt Collection, courtesy of the Wadsworth Atheneum, Hartford, Connecticut.

Colonel Hiram Berdan (1823–1893), the American inventor of several improvements in small arms and ammunition, including breechloading mechanisms modified by Russian officers and adopted by the Russian army. Photograph from a series by the Matthew Brady studio, c. 1864, of Union officers, courtesy of the Library of Congress.

Colt's Patent Fire Arms Manufacturing Company. Colonel Alexander Gorlov used the Colt factory as a research and design facility to test the many new breechloading rifle systems available after the American Civil War. Russian ordnance inspectors worked here when the Colt factory began making the Berdan 1 infantry rifle for the Russian government in 1868. Lithograph printed by L. Schierholz, n.d., courtesy of the Library of Congress.

Colt's Patent Fire Arms Manufacturing Company, Hartford, Connecticut. In the foreground is the Connecticut River. The original building, built between 1854 and 1855, was destroyed by fire in February 1864. The onion-shaped dome was added after Samuel Colt's first trip to Russia. Sky-blue with gold stars and illuminated at night, it is still a striking feature of the Hartford skyline. The charging colt, however, is no longer perched at the top. Lithograph by E. B. and E. C. Kellogg, in *Proceedings at the dedication of Charter Oak Hall* (J. Deane Alden, 1866), courtesy of the Library of Congress.

The Smith and Wesson factory in Springfield, Massachusetts. Colonel Alexander Gorlov and other Russian officers worked here after Smith and Wesson and the Russian government contracted in 1871 for 20,000 Model 3 revolvers. Photograph published by the Detroit Publishing Company, c. 1900–1910, courtesy of the Library of Congress.

The Tula Armory, on the banks of the Tulitsa River. Founded by Peter the Great in 1712, the armory was expanded many times and rebuilt after a fire in 1835. Work was centralized and mechanized in 1873 after the armory began to make the Berdan 2 infantry rifle with machinery purchased in England. Lithograph by Veierman from a photograph by Brenke, in *Niva* 41 (1877): 661, courtesy of the Library of Congress.

The Izhevsk Armory on the banks of the Izh River in the Ural Mountains. The armory opened as a small arms factory in 1808 and was mechanized in the early 1870s to make the Berdan 2 infantry rifle. Lithograph by M. Rashevskii from a photograph by Nikitin, in *Niva* 43 (1887): 1053, courtesy of the Library of Congress.

Master armorers of the Izhevsk Armory wearing ceremonial tunics. Each armorer who distinguished himself in skill or in conduct received a ceremonial tunic (*paradnyi kaftan*) bearing the name of his gunmaker's shop. Lithograph by Fliugel' from a photograph by Nikitin, in *Niva* 43 (1887): 1072, courtesy of the Library of Congress.

received the rights and privileges of, although not membership in, the *meshchane* estate, a legal category of small producers and traders living in towns. They were exempt from billeting troops in their homes. They could establish their own workshops, engage in other trades, and leave home to sell their wares. (The concentration of samovar and cutlery works in Tula came about because of this latter privilege.) They could purchase raw materials (iron and coal) from the state at wholesale prices; that is, at the same prices paid by the armory. Finally, they could not be required to do unskilled labor by the armory.[18]

In legal matters the Tula gunmakers fell under three jurisdictions. In domestic affairs, the armorers were tried by the small claims court and by the factory police; for civil offenses by the guild council; and for criminal offenses by a military court. The administration of the small arms factories could at any time curtail "outside work" and oblige the gunmakers to work solely on state orders. The result was a caste, rather small in number: in 1861, the three government armories employed 8,353 gunmakers and 1,869 apprentices.[19]

At the same time, obligations and restrictions bound the caste of gunsmiths to the gun factories and to the state. Until 1864 gunsmiths preserved their own corporate organization and property. In theory they were geographically as well as legally separated from the surrounding population. The Tula gunmakers, for example, lived on the left bank of the Tulitsa River in a separate settlement, a residence forbidden to outsiders. Like the inhabitants of most of Russia's settlements, the gunmakers did not always live in the armorers' quarters; nevertheless the settlement remained as an administrative entity, if not a residential or occupational one. For a long time, the gunmakers were forbidden to sell arms except through official channels, giving the government a virtual monopoly in the consumption of their product. Overwhelming constraints on mobility also perpetuated the closed corporation. The ascribed status was passed on to children; sons of gunmakers were required to learn the trade of their fathers. Assigned the legal status of "factory inhabitants," gunmakers at Izhevsk and Sestroretsk were forbidden to transfer to another estate. Although in theory the Tula gunmakers could transfer at any time with government permission, in practice this was rarely granted.[20]

Given the prevailing labor system of serfdom and the ideal, if not always the reality, of the service state, the introduction of a modicum of market relations and labor mobility into Russian society was premature. Nevertheless, the seeds of legislation to permit both market and mobility were planted during the first half of the nineteenth century, well before the famous emancipation edict of 1861. Regarding the possessional serfs, for example, in 1831 the Ministry of Finance could

propose the transfer of individual possessional workers to the merchant or *meshchane* estates. Four years later, possessional factory
owners were accorded the right to dismiss possessional workers provided that such dismissals did not diminish the factory's output.[21]
Were there similar chinks in the armor of indenture at the state small
arms factories?

Since the reign of Catherine II, there had been discussion of allowing
surplus Tula gunmakers to enter the merchant or *meshchane* estate.
In 1823 a commission to investigate ways of improving the condition
of the Tula gunmakers observed that a large number of gunmakers
were reduced to idleness and pauperism due to a lack of state orders.
More than a decade later the commission recommended that the size
of the gunmakers' corporation be reduced from an estimated 7,364
to only 2,200 by allowing gunmakers to leave the corporation.[22]

Arguments for keeping the Tula gunmakers under the authority of
the armory, voiced by Count Arakcheev and others, prevailed. The
government regarded the surplus number of gunmakers to be desirable
in times of urgent need for arms. The example most frequently cited
was the patriotic response of the Tula armorers to the appeal of the
sovereign in 1812. During the next three years, Tula delivered five
hundred thousand arms to the nation. Whether the situation remained
the same half a century later with escalating arms needs and rapidly
changing technologies was apparently not considered. But there may
have been more important reasons why the state thought this system
advantageous. The government assumed that if the gunmakers were
allowed to enroll in another estate, the best gunmakers would be the
first to leave, thereby damaging the capabilities of the corporation as
a whole. Indeed, gunmakers who did leave, albeit temporarily, to engage in other trades were obliged to pay an absentee tax. Although
this tax allegedly helped support the corporation as a whole—and as
such was no more than a welfare payment—its major effect was to
discourage long-term departures. The 1823 commission realized that
retaining gunmakers who wanted to leave did not make the corporation as a whole better off. This apparently was not the prevailing opinion of the commanders, who had to grant permission to any gunmaker
seeking to enroll in another estate. Concern that the departure of a
skilled worker would jeopardize the corporation, and hence Russia's
needs in arms, would suffice to refuse the request. It seems that in
any case the issue was moot; few armorers were able to accumulate
the capital necesary to enroll in the *meshchane*, let alone the merchant, estate, and sources do not suggest that the factory administration was inundated with requests by gunmakers to leave their corporation.[23]

At the same time the government was considering loosening its hold over the Tula gunmakers, it tightened its grip on those at Sestroretsk. There gunmakers had been regarded as serving military duty; according to an 1807 statute, the length of service was thirty years, after which the gunmakers were no longer obliged to the armory. However, in 1823 a statute established that gunmakers were permanent factory inhabitants and permanently obliged to the armory. Ten years later another statute ascribed to the gunmakers the same legal status as soldiers and put all work at the armory on a military regimen.[24]

That both skilled gunmakers and other laborers were indentured to the armories suggests several features of the Russian firearms industry during the century and a half preceding the Crimean War. First, the number of gunmakers was not large for Europe's most populous nation; nor was there a large number of gunmakers in the private sector who would complement the government armorers. Nevertheless, there was an adequate number of gunmakers to fill state needs during a period of modest increases in weapons needs. Second, although the numbers varied over time and the available estimates are not altogether consistent, it would appear that only one-half or less of the total work force at the armories consisted of gunmakers and apprentices. The relatively large number of semiskilled and unskilled laborers suggests technological backwardness in the armories.[25] Finally, the labor force was largely immobile. Labor, unskilled and skilled, was neither purchased nor organized by the market but commandeered from within, in this case within the armories themselves.

Recruitment and Training

Closed corporation meant that no one from other estates could enter the corporation of gunmakers. Yet some means had to be devised to recruit new gunmakers. Unlike other government factories that recruited free labor, but like the possessional factories, the armories forcibly assigned workers. Over the years the state recruited for the corporation of armorers from three sources. Throughout most of the eighteenth century, the government armories relied on a labor draft of the peasant population, typically by transfers of peasants from other state factories to the armories or by purchase of privately owned serfs. Thus, for example, the first Sestroretsk gunmakers were transferred from the Olonets state factory; 457 were transferred in 1724 alone. Later, Tula gunmakers were also transferred to Sestroretsk. In 1763 the core of armorers at Izhevsk was purchased from Count Shuvalov; in 1820 Tula purchased 350 serfs from Naryshkin, and Sestroretsk purchased a large number of serfs from Count Saltykov.[26] Army recruits were a second important source of gunmakers, especially at Izhevsk and Sestroretsk. When the Izhevsk iron foundry was converted in 1807 into an arms factory, 700 recruits were assigned. (More than 100 foreign craftsmen also were recruited when the Tula commander refused to transfer any of his skilled gunmakers.) At Sestroretsk 1,206 army recruits were brought in from 1808 to 1810 alone.[27] Although effective in providing a massive short-term infusion of laborers, this method yielded unskilled gunmakers requiring on-the-job training. Finally, and increasingly in the first half of the nineteenth century, sons of gunmakers themselves became apprentices. By 1861, there were 904 apprentices at Izhevsk, 839 at Tula, and 126 at Sestroretsk. Sometimes sons of peasants assigned to the armories also became apprentices. Cancellation of the providing household's obligations to the factory suggests that the government felt it needed to offer an enticement to recruit apprentices. These apprentices learned to read and write and, in theory, learned their craft at the armory. However, the statutory obligation of boys to learn the craft of their fathers was frequently not enforced, partly because many fathers themselves, especially at Tula, as we shall see, did not practice their craft.[28]

As elsewhere, vocational schools run by the factories opened only during the second half of the nineteenth century. Thus, as might be expected, the armories trained army recruits, purchased peasants, and apprentices on the job. Only in the case of peasants purchased from private manufactories, such as the Shuvalov peasants purchased by Izhevsk, or transferred from state manufactories, such as the Olonets peasants transferred to Sestroretsk, were the armories likely to get a skilled labor force.[29] Significantly, one form of training is seldom encountered in the sources. There is little evidence that gunmakers went from one armory to another or to a private factory to learn, or to teach, a new technique. This suggests that this important form of training and of technological diffusion either occurred so frequently that it seemed unremarkable to contemporaries or occurred very rarely.[30] We will return to the point below; suffice it to say here, the latter seems more likely.

Wages and Conditions

Various statutes determined the nature of work, the work regimen, the amount of wages and provisions, the length of the working day, and the number of holidays for workers in all state factories. During the period between 1845 and 1849 a Tula gunmaker working full time on government orders earned an average of twenty-four rubles per year. Sestroretsk barrel forgers, in contrast, earned more than 120 rubles annually. However, Sestroretsk gunmakers had to pay for their own raw materials and provisions; moreover, they had little outside work to supplement their income.[31]

Although gunmakers were considered to be well paid, structural factors served to lower real wages. The most onerous of these factors, and a bane of Russian workers in all branches of industry, was a myriad of fines and deductions. References to deductions for defects in gun parts are encountered repeatedly in the sources. Deductions for defects were taken out of piece rates, and in years when the percentage of defects was high (up to 80 percent), most piece rates were eaten up by such deductions. This was also the source of a dispute between the Artillery Department, which claimed that defects originated in poor materials, and the Department of Mines, which claimed that the defects originated in poor workmanship. The dispute, never satisfactorily resolved, figured in the sense of urgency behind technological innovation during the Era of the Great Reforms.[32]

As elsewhere during the first half of the nineteenth century, wages for skilled gunmakers were increasingly based on piece rates. At Sestroretsk, for example, an 1823 statute dividing workers into gunmakers and mechanics stipulated that while the latter were to be paid a fixed hourly wage, the former were to be paid by the piece.[33] Those defending this system argued that gunmakers would feel a greater sense of responsibility for their work and factory managers could exert more authority and quality control. Hourly wages paid by the week allegedly gave little incentive to the gunmaker to work quickly and carefully. According to one historian of the government armories, piece rates were paid only two or three times annually.[34] Such infrequent payments contributed to the indebtedness of the armorers and effectively disassociated efforts and rewards in the minds of gunmakers, thereby cancelling the major rationale of piece rates.

The small arms industry provides an excellent example of government control in industry and labor policy; it also provides an example of a complementary policy, government paternalism. Such a policy of paternalism not only mitigated the policy on wages but also shielded gunmakers from the exigencies of fate or of the market. The opportunity to work on side orders, especially at Tula, improved the armorers' condition. Izhevsk gunmakers were allowed to earn supplemental income from truck gardens and livestock. Moreover, a variety of subsidies and relief payments, described often in the sources, supplemented wages. Land was granted to gunmakers free of charge for dwellings and, at Tula, for private workshops. Gunmakers were also granted access to state forests for building materials and fuel. The factory provided a hospital, school, and an almshouse. For example, from the mid-eighteenth century on, Sestroretsk had a clinic with fifty beds. Wages were supplemented by provisions such as rye flour, and in years of bad harvests flour was sold at wholesale prices from factory stores. Gunmakers received allowances for children, and gunmakers who had worked twenty-five years received a pension. An armorer who distinguished himself in skill or in conduct was rewarded with a ceremonial tunic identifying his gunmaker's shop. Finally, although the gunmakers worked ten to twelve hours per day, six days per week, they were allowed one week off for fasting and two weeks off for mowing. Their total working days per year came to only 270, a common, although baneful from the management's point of view, feature of Russian labor.[35]

Government paternalism went beyond policies of relief and subsidies. The culture of preindustrial labor everywhere has generally recognized some sense of a just wage; in Russia, government and people were held together by a mutual recognition of a just job. The government granted to the workers the right to an adequate subsistence. State policy perpetuating the closed corporation removed competition. But perhaps the most important aspect of this governmental paternalism was the provision of job security. In this sense the gunmakers were like Russia's possessional factory workers. Regardless of business conditions, a possessional worker received a constant wage. Although factory owners could curtail production, they were obliged to provide work and to pay wages even when it might have been more profitable to lay workers off.[36] It is likely that job security, and the concomitant preservation of the social order, was what Arakcheev had in mind in recommending that gunmakers remain tied to the factory. According to the 1838 Code of Military Regulations, the foreman was supposed to make sure that gunmakers were not "punished by either the absence or the inadequacy of work."[37] (This regulation could be compared to the 1823 recommendation that only 2,200 of more than 7,000 gunmakers be employed on state orders.) The stated purpose of M. Subbotkin's history of Izhevsk was to describe the changes necessary to improve manufacture of arms and to better provide for workers. As late as 1870, a commission appointed to study the revamping of the Tula armory was told to investigate ways to mechanize the factory without throwing gunmakers out of work.[38]

One might suppose that secure workers would be better, more satisfied workers. However, studies by commissions in 1823 and 1849 found depressed conditions, especially at the Tula armory. State orders were inadequate and private workshops were idle; few gunmakers had a horse or cow; and the 1849 study showed that two-thirds of gunmakers' homes were in need of repair. Perhaps most debilitating to gunmakers was a system of collective responsibility, generally associated with the Russian peasantry but characteristic of all Russian life. The entire society of gunsmiths was collectively responsible for raw materials alloted to the various shops as well as for the finished products. Better-skilled or more capable gunsmiths were responsible for the rest, a system that allegedly bred apathy, sloth, dishonesty, indifference to work, and the absence of self-improvement. As a result, although they had certain privileges and immunities, in their overall material and moral condition, according to the report of the 1823 commission, gunmakers more nearly resembled paupers than proud craftsmen.[39] Similarly, a history of the Sestroretsk armory concluded that the frequency of defects demonstrated poor knowledge of gunmaking. The skilled gun craftsmen in Russia, fewer there than in Europe, apparently lacked even the craft pride of their European counterparts.[40]

According to an 1850 study of the Tula armory, less than half of the total number of gunmakers were listed as actually working in the armory.[41] The remainder did contract work for the armory in small private workshops.

The Organization of Work

At all armories work was organized by shop and work team, and the multilayered authorities were necessary to transfer work from one division to another. The division of work by the chief engineer reflected the static world of the closed corporations of gunmakers. True to the collectivist culture of preindustrial labor in general and of Russian labor in particular, jobs were accepted not by individual gunmakers but by entire shops, typically with the involvement of as many workers as possible. As a visiting British engineer observed, "Throughout the whole of Russian [iron] works one is struck with the large number of people congregated together to do the work which could be done equally as well by a less number."[42] Work was divided into work-weeks according to the number of gunmakers available. A sudden increase in work could be accomplished, in theory, only by retaining a surplus number of gunmakers in the armory. After being assigned a weekly quota of work, each shop steward received the proper amount of supplies from the factory storehouse and distributed them to the bosses of each work team, who in turn distributed them to the individual master gunmakers. At the end of each week the shop stewards collected from work teams the finished products, registered them, and then delivered them to the steward of the next shop, who apportioned out the product to his gunmakers, continuing the cycle. Defective work was returned to the work team or shop responsible. In this way preliminary inspection of parts preceded final inspection and proof of each weapon. Each gun passed from work team to work team and from shop to shop until it was ready for final fitting, inspection, and proof, jobs performed exclusively at the armory, even at Tula. Guns that passed inspection were submitted to the factory arsenal where they awaited delivery to the national arsenals (St. Petersburg, Kiev, Briansk) or procurement by regiments.[43]

At all three armories work was subcontracted to various shops responsible for submitting all or parts of a weapon. At Sestroretsk and Izhevsk, most shops were on the factory premises. At Izhevsk, for example, the work on the guns was performed in fourteen stone buildings in one unified factory. Gunmakers did not own their own shops.[44] At the larger Tula armory, in contrast, much of the work was distributed to gunmakers who had their own workshops and tools. These gunmakers acted as subcontractors and in turn hired other gunmakers. An 1849 commission formed to investigate the reasons for the decline of the armories and for generally poor conditions for gunmakers found that while 1,260 Tula gunmakers worked at the factory, 2,225 worked at home; twelve years later 61 percent of 3,916 gunmakers were still working at home. All finishing and fitting operations on the locks, for example, were done in individual workshops. An 1865 survey counted 216 workshops, only five of which employed more than 15 workers.[45] For more than a century and a half, the center of gunmaking in Tula was the armorers' settlement, not the factory. According to one English visitor,

> It surprised us to find, that instead of having one large establishment, where all branches of gunmaking could be prosecuted together and where all workmen could carry on their various departments under proper inspection, nearly all the work is performed by the blacksmiths at their own houses. When one has done to a musket all that belongs to his branch, it is sent away to another, and so on till it has traversed Tula a dozen of times. We thus found hundreds of blacksmiths carrying about muskets from place to place when labouring hours were over.[46]

Given the government's national security needs, the closed corporate nature of the gunmakers, the absence of market conditions in labor, and state paternalist attitudes, surplus gunmakers had to be employed and provided for. One of the privileges granted to gunmakers was the right to take outside jobs, such as the manufacture of utensils, axes, iron wares, and samovars, to supplement their income. Many Tula gunmakers also engaged in repair work and retailing, which frequently caused absences from Tula for months at a time.[47] Outside jobs were not limited to the relatively decentralized Tula gunmakers. Many Sestroretsk gunmakers, who worked almost exclusively in the small arms factory, also took outside orders. According to several accounts, the commander, whose permission was necessary for such work, saw to it that as many gunmakers as possible were involved in outside jobs.[48] The commander knew that at any time he could recall gunmakers working on outside jobs and force them to commence work on an armory order. More important, outside jobs provided gainful employment for the gunmakers.

As a result of this system, gunmakers engaged in a variety of jobs, at times only incidentally practicing the gunmakers' craft. It would appear that although the factory system may have reduced such occupational diversity, it did not altogether eliminate it. Although engaging in a variety of outside jobs to supplement income is a feature most commonly associated with the Russian peasantry, this occupational diversity also characterized one of the most highly skilled crafts.

Outside jobs posed a greater risk to gunmakers than did government orders. Did the competition fostered among gunmakers to win outside contracts foster competitiveness and efficiency in the fulfillment of government orders? According to some contemporaries, the market for outside work was smaller at Izhevsk and Sestroretsk—at the former because of its isolation and at the latter because of its poor competitive position relative to the metal shops of nearby St. Petersburg—and the Izhevsk and Sestroretsk gunmakers had less need for and were less interested in outside jobs than were the Tula gunmakers. Lacking the means, the independence, the leisure, and the spirit of enterprise, the Sestroretsk armorers could not produce either high quality or low cost wares.[49] The Tula gunmakers allegedly had a greater keenness for outside work and a greater sense of competition. According to F. Graf, it was clear that the gunmaker with greater skills could make more money on outside jobs. Graf concluded in 1861, "Competition is the most energetic stimulus, and it is only by means of competition that new inventions can be applied to industry.[50]

The small, dispersed workshops were blamed for, among other things, the alarming number of barrel defects in Russian small arms. According to M. Subbotkin, an officer at Izhevsk, there were days when up to 90 percent of the work was unfit. According to one study of Tula, independent barrel makers had primitive ovens consisting of little more than a hole in the ground. A hole in the ceiling served as the chimney, and primitive leather bellows were raised and lowered by hand. Independent barrel makers preferred to economize on charcoal. Under such conditions it was difficult to heat the iron to a sufficiently high temperature. The study concluded that it would be preferable to forge barrels in the factory rather than in the houses of gunmakers.[51]

Guns produced by the state arms factories, although on occasion impressive, generally did not match the quality of guns produced by Liège, Birmingham, or London. The description of state armories by an English observer reiterates several shortcomings already in the record:

> Percussion locks, of course, here, as everywhere else, are fast driving all others out of use. Little care, however, is exercised in selection of metal for barrels. . . . The boring process, as well as the proving, are also very roughly conducted. The consequence of all of which is, that accidents are of frequent occurrence from these guns; and, perhaps, will continue to be so, till a more general diffusion of taste for field-sports encourage manufacturers to produce a superior article. . . . Generally speaking, we did not see much to admire in Toula workmanship. The things are very slight and of inferior finish. When used for awhile, the joints are always going wrong, and screws are never a week fit for use. Except the snuff-boxes, few fancy articles receive the labour that would be bestowed on toys in England. Heavy things, however—water-pipes, fittings for furniture, etc.—are substantially done.[52]

The shop jobs, the subcontracting system, and the supplemental jobs all fostered a decentralized organization of production, and contemporaries were divided over whether the system was good or bad for the Russian firearms industry. From the point of view of the government, the workshop system with surplus gunmakers earning at least part of their livelihood elsewhere resulted in considerable savings to the treasury. According to one estimate, the government saved one million rubles through the workshop system at a time when total capital investment in the factory was only two million rubles. In addition, "worker satisfaction," if a modern term may be permitted, was frequently cited as a defense of the decentralized workshop system; the gunmakers themselves allegedly preferred working at home with their families close by.[53] Repeated efforts to centralize the Tula armory during the eighteenth and nineteenth centuries were thwarted by the armorers, the government, or a collusion of the two. Attempts to centralize armories demonstrate important ways in which the interests of government and labor complemented each other.

An early attempt to centralize work failed soon after the Tula factory was built in the eighteenth century. Peter's decree establishing the state armory declared, "In order to turn arms production in Tula to good account and hasten it, an arms factory is to be built, where all artisans will be making guns unceasingly; the making of guns in the homes of artisans shall cease henceforth." A subsequent prohibition on milling and fitting operations in private workshops was resisted by the armorers and was never enforced.[54] A century later, the government considered a more serious attempt at centralization. In 1824 engineers at the armory began drawing up a design for a new factory. A major design consideration was the amount and type of jobs to be performed at the factory itself. A proposal for a factory concentrating all work in one location was supported by the armory's commander, General Staden. By eliminating work in private workshops, Staden argued, "all work will be performed under constant supervision and, consequently, it will be better. . . . With the gunmakers at all times in sight, supervision of their conduct will be easier and will not slacken."[55] Although it is tempting to assume that the Russian government would place the highest value on the supervision of its subjects, such a degree of centralization was deemed too costly. By permitting much of the work to be done at private workshops, the treasury saved four million rubles in new buildings, and in 1827 Nicholas I approved a design for a new factory in which only the most important fitting and assembly tasks were performed in a central location.[56]

A decade later the question reappeared when fire destroyed the factory buildings. The opportunity to centralize production again arose. After the fire the argument against centralized production was even greater: rebuilding the armory was costly and concentrating production in one place, as the fire demonstrated, was risky. Furthermore, the government argued, the European small arms industry, especially in England and Belgium, was decentralized. Consequently, Nicholas I again ordered that most gun operations continue in private workshops, where armorers could use their own tools and equipment; only the most important finishing and assembly tasks were to be performed at the central factory. Of 3,074,291 rubles appropriated for factory construction after the fire, only 65,917 rubles, or 2 percent, were appropriated for machinery; of this, 50,907 rubles were appropriated for repair of existing machines. In 1847 the question of eliminating work in private workshops again was raised, but a proposal to build more work space on the factory grounds was rejected as being too costly.[57] There the matter stood until the rearmament drive of the 1860s and 1870s.

Two additional points must be made about the organization of work at the armories. First, differences among armory practices and recruiting methods did not yield a favorable conjuncture for the Russian small arms industry. As is evident, recruiting peasants and army conscripts did not produce skilled gunmakers. The army recruits, in particular, were illiterate, unskilled, and, most important, not interested in the results of their labor. What appeared at first glance to be a source of cheap labor in the end produced high-cost results. Izhevsk and Sestroretsk, the most centralized armories, commonly employed these conscripts. It was generally assumed that apprentices learning on the job yielded a more motivated labor force at less cost. Tula, the most decentralized armory, used this form of recruitment. Thus the more centralized, modern armories trained an unskilled and unmotivated labor force at high cost; the more decentralized armory trained a more motivated labor force at low cost.

Second, gunmakers at Tula were considered the most competitive for outside contracts and therefore the most skilled. More competitive and skilled, the Tula armorers would presumably have been in a better position to innovate. Yet they worked largely at their own small workshops. Thus, even had the Tula gunmakers been innovative, the Tula armory would not have reaped the benefits. Working largely at a centralized factory, the Sestroretsk workers would have been in a better position to innovate at the factory. But, as has been observed, they were less competitive and therefore allegedly less skilled and innovative. In short, the best trained gunmakers worked at the most archaic factory, less conducive to innovation and less subjected to central supervision.

The closed corporation of gunmakers, the shop organization of work, the considerable decentralization, especially at Tula, and the relatively large number of semiskilled and unskilled laborers employed at the armories suggest technological backwardness. Well before the second half of the nineteenth century, Russia did have a system of planned production and procurement and, at Izhevsk and Sestroretsk, of more or less centralized government armories. And, as Nikolai Leskov demonstrates, the Tula craftsmen were capable of shoeing the toy steel flea. Clearly, there were advances in metalworking technology at the three government armories in the prereform period. Did Russian armories keep pace with the innovations then being introduced in the West? Specifically, did the Russian armories exhibit a trend toward general mechanization, a division of labor, the sequential application of special-purpose machines, and uniformity of parts?

Innovation

Peter's decree establishing the Tula armory in 1712 included an order "for the best method in arms-making, to build machines in the appropriate place at this arms site so that guns and pistols can be drilled and ground, and broad swords and knives sharpened with water." Both prerevolutionary and Soviet historians have documented early examples of the application of water power in the metalworking trades. According to an 1826 account of the Tula armory, water-driven lathes and boring machines, in use as early as 1714, were so sturdy that they were still in use a century later, a somewhat dubious distinction.[58]

Although there are several indications that Russian industry utilized a variety of machines, the first machine-building factory was not founded until 1790. The duty-free import of machines in the early nineteenth century, coupled with the prohibitively high tariff on imported pig iron, impeded the growth of a native machine-building industry. In 1855, Tegoborski wrote,

> We cannot be blind to the fact that most of our iron-masters, reposing on the cushion of protection, long neglected to follow the progress of this industry in foreign countries: any improvements introduced have been very recent and very exceptional. . . . The admission of foreign iron would give a great impulse to the home construction of machinery.[59]

According to the Ministry of Finance, in 1850 Russia had only twenty-five private machine plants employing only 1,475 workers, the products of which were valued at 424,000 rubles; by contrast the value of imported machines, tools, and instruments was more than three times greater. Accordingly, until the 1850s, machine-building was developed chiefly at government factories.[60]

Early use of machinery facilitated three innovations: the division of labor, the introduction of special-purpose machines, and fabrication of uniform gun parts. All three innovations have been documented in an 1826 study of the Tula armory, a startling account of the advances in the application of special purpose machines and in the attainment of uniformity. Rich with illustrations of machinery and descriptions of fabrication techniques, the study was deemed important enough to submit as evidence of Russian production methods before the British Parliamentary Select Committee on Small Arms in 1854.

Iosif Khristianovich Gamel' opened his study of the Tula armory with the arresting claim that recent mechanical improvements had surpassed those of even the best English small arms factories and that "mechanically no other small arms factory in the world can be compared to it."[61] The improvements mentioned by Gamel' were introduced by an English mechanic from the Birmingham firm of James and Jones. Because the British government did not show interest in purchasing the gunmaking machinery patented by the firm in 1811, James and Jones approached the Russian government. In 1817 John Jones (c. 1786–1835) arrived in Tula to help install lathes from

Birmingham and to act as chief of the mechanics' shop. Working much like the inside contractors of American armories, Jones developed several lathes and stamping, die forging, and milling machines to solve specific production problems at Tula. According to Afremov's history of the armory, Jones lived in the armorers' quarters for seventeen years until his death shortly after the fire of 1834 that destroyed many years of work.[62] The improvements illustrate the more efficient use of raw materials, the division of manufacturing operations into discrete parts, the application of special-purpose machines, and the attempt to make uniform parts.

The forging of barrel blocks provides a good example of more efficient use of materials. Barrel blocks, which had formerly been forged by hand at home, were now forged at the armory in six furnaces, especially built for Jones's shop. According to Gamel', the amount of metal used was reduced by 40 percent, the work went "easier and quicker," and "the entire foundry shop benefits from his method."[63]

Jones divided manufacturing operations into discrete parts, which resulted in a considerable division of labor. According to Gamel', fabrication of firearms was divided into thirty-one separate operations:

> A total division of labor exists in all sections of the plant, such that one workman always performs the same operation. I doubt that any factory in the world follows the system so much praised by Adam Smith more than is practiced at the Tula small arms factory. (The infantry musket passes through more than one hundred operations.)[64]

This division of labor may also have been facilitated by the existence of work teams in various gun shops. The entire work team was responsible for the finished product; the individual gunmaker was responsible for only a specific part. However, it is not clear whether this promoted innovation, one of the chief ends of the division of labor, according to Adam Smith.

Special-purpose machines also came from Birmingham. Such machines, of course, were not completely new in Russian armories. Soviet historians have documented the use of special boring and grinding machines in the eighteenth century. A late eighteenth-century machine could produce four barrels simultaneously, thus eliminating the most labor-intensive manual operations. Pavel Zakhava, a mechanic who came to the armory in 1810, invented a machine that could bore twenty barrels a day and that "far exceeded anything in America or England for that purpose"; in 1812 Zakhava built a special lathe for turning barrels. Britkin concludes his study with praise for the native machine-builders:

The wide application of the principle where the design of a long line of specialized machinery is based on a number of standard components and subassemblies was a procedure not adopted anywhere else; so were standardization and interchangeability. Replacing labor-consuming operations by special highly productive machinery with power drives and by semi-automatic machinery, which both Batishchev and Zakhava attempted, is the guiding principle of twentieth-century machine building.[65]

Jones is most often credited for the invention or application of several special-purpose machines. A "superb machine" Jones invented for turning barrels permitted "far fewer cracks to show up at the proofs."[66] From 1818 to 1821, he replaced hand-forging with mechanized die-forging of the parts of the lock. All components of the lock except the spring and priming pan, formerly produced at home without precision, were now standardized and made in one shop by special-purpose machines. As a result, "the parts come out of the die in a finished state" and "the fitters actually have little else to do but file and polish the parts."[67] By 1825, 144 machines, including several special-purpose machines for milling and filing, were installed. According to Gamel', the "superb" machines Jones invented for die-forging the locks and turning barrels would

immortalize his name in the history of not only the Tula factory but also of the gunmaking craft. . . . It is unlikely that anyone in England could make a better turning lathe than Jones. . . . The shop where Jones's dies and presses are now located is indeed a pleasant sight and is one of the most interesting divisions of the entire plant.[68]

There are also indications that uniformity of gun parts was becoming a goal of production in the Russian small arms industry. An order of the War College dated 1798 is perhaps the first indication of the goal of uniformity. The gunmakers were ordered to make gun parts without "the slightest difference in size, in weight, or in proportion." All parts were to be even so that they could be replaced in the regiments in case of damage. Such standardization of parts was to be achieved by precision gauges. One year later the Sestroretsk armory was making three hundred arms for the Preobrazhenskii regiment using interchangeable parts. The Soviet history of Sestroretsk concluded, "The production of interchangeable parts was a great achievement of the Russian craftsmen, accomplished by utilizing machine technology."[69]

A similar claim has been made for Izhèvsk. According to the Soviet historian A. A. Aleksandrov, the first commander of the small arms factory, A. Deriabin, reported to the Ministry of Finance in

1809 that the locks made by the Izhevsk gunmakers were greatly improved:

> The soldier can take apart and assemble this kind of lock in one minute without the slightest difficulty. Only a screwdriver is needed. All lock parts are affixed on only one screw. Among the most important exterior improvements is the production of all parts of the gun of uniform size with such complete precision so that any part of one gun will fit another gun.[70]

But, again, the most startling accounts of uniformity pertain to the efforts of the mechanic from Birmingham. Before 1819 the forging of different lock parts was performed manually by the gunmakers in their small workshops. As a result, parts were not accurately forged and needed to be fit together by the tedious process of filing. According to Gamel', Jones's use of dies "easily produced" parts that required almost no filing and were "almost finished and perfectly uniform":

> Jones attained the required uniformity of lock parts to a much greater degree than ever before. In other countries, uniformity of lock parts has been deemed impossible even after many lengthy attempts. . . . Of course, even with the most advanced mechanisms, the lock parts may turn out to be not exactly uniform if the workmen are not assiduous.[71]

Even more striking was the degree of interchangeability allegedly achieved by Jones's methods at Tula. Using Jones's dies, the lock parts "need only a very small effort to be converted to perfectly finished interchangeable parts." The common method to demonstrate interchangeability at that time was to take a batch of muskets, disassemble and scramble their parts, and then reassemble the muskets from randomly selected parts. Gamel' claims that he once saw at the Tula armory a large

> number of locks which had been disassembled in the inspection shop and whose parts had been mixed. When these parts were reassembled, they fit together so precisely, it was as if they had been deliberately fitted. This conclusively shows the superiority of the methods Jones has introduced at Tula to make locks.[72]

Gamel' was not the only witness of such impressive demonstrations. Nicholas I, visiting the Tula armory in 1826 with his brother and General of Ordnance, Grand Duke Mikhail Pavlovich, picked thirty muskets from the arsenal to be disassembled. On this occasion, lock parts as well as entire muskets were reassembled, illustrating "complete uniformity." In 1837, the heir to the throne, Alexander

Nikolaevich, visited Izhevsk and also witnessed the customary scrambling of rifle parts to demonstrate interchangeability.[73]

Other testimony gives credence to the account by Gamel'. Although a study of Russian arms manufacturing was not on the agenda of the British Parliamentary Select Committee on Small Arms in 1854, incidental testimony from British observers suggests that wide use of special-purpose machine tools and of stamping operations enabled uniformity of parts. James Nasmyth and Richard Prosser, a patentee of machinery for making nails, buttons, and tubes, both praised the machines and production at Tula. Describing improvements in the stamping of wrought iron, Nasmyth revealed that he had been supplying steam hammers to Russia, where his improvements "had been carried out with the greatest success." Nasmyth believed that the Tula armory made most of its muskets by machinery, specimens of which he exhibited to the select committee: "To show the integrity of each part, they have sent me duplicates just as they came from the stamp, and others left with a little frill, not filed; they have been struck; I believe they have the same process at Birmingham, but I think it is done in a ruder manner."[74] Later, Nasmyth admitted that his machines were not responsible for the finishing at Tula:

> No, [the machine] was made at Tula by the Russians; but they are so well satisfied with the results of it, that the Government have determined to extend it very considerably, and hence the demand for the estimate which we handed in of 12,000, which was to be mere reduplications and extensions of those machines that they have already at Tula. I must say that they are very admireable contrived; they were submitted to me to modify where I thought fit, but there was very little for me to suggest. . . . The parts were finished in the best sense, what a mechanic would call the best kind of finish.[75]

Prosser displayed for the committee the plates and description of Tula machinery found in the study by Gamel' and testified, "I do not think that there is any part of the gun except the stock, but what is in some way operated on by means of labor-saving machines or tools." Prosser believed that the machines and tools were made in Russia and were still in use, and that an English mechanic (presumably Jones) was still at Tula. Moreover, Russia, he thought, had machinery that England did not. Finally, Prosser repeated the familiar tale of musket assembly from disassembled parts:

> The result of that machinery is this, that about 1822, Mr. Fairy, an engineer of London, went to Russia and at Tula he saw this machinery in operation.

It was some time about 1817 that . . . he saw there twelve Russian soldiers come with muskets made by this machinery; there were twelve baskets before them, into which they put the stocks, locks, and barrels and everything else, and then fell into rank again. Each man then went to the baskets and took up a stock, a barrel, and everything and put them together, and fired them off in two minutes.[76]

Perhaps most intriguing, both Nasmyth and Prosser suggested a similarity between the Tula machinery and that of the Colt factory. Nasmyth testified that "the manner in which they [the Tula machines] were designed was admirable, and very much on the same principle as those of Col. Colt." Prosser described Colt's machinery as being "very much like the Russian plan." Ironically, the "Russian plan" may have been synonymous with the manufacture of uniform parts. Nasmyth was uncertain whether uniformity was attained at Tula. He believed that "those parts that come from the machine would be fit to put together in the state they are" but was unsure whether they were "so put together."[77] However, as indicated earlier, he was certain that Colt had not attained even this degree of uniformity.

Thus Jones apparently brought about more efficient use of materials, a significant degree of mechanization, the application of special-purpose machines, the reduction of finishing operations, and a noteworthy degree of uniformity and interchangeability. But Jones's contribution went beyond the invention of special-purpose machines. In several shops, Jones organized production, helped train the work force, and facilitated the division of labor. When the workers in the barrel forging shop "saw that at the first proof, not a single barrel broke, all the forgers asked him to teach them his method."[78] When Nicholas I during his visit to the armory asked Jones his opinion of the technical competence of the Tula gunmakers, Jones answered that, although at first he was given workers who knew nothing, now they were just as skilled at making tools and machines as the best gunmakers in England.[79]

There is evidence that Jones's activity was duplicated elsewhere, at least for short periods. In his introduction, Gamel' expressed the hope that the description of Jones's machines would be useful to Russian manufacturers, "since many of these machines can be put to good use in many other works and factories." According to a study of the Ministry of Finance, Jones taught local metalsmiths new methods of shaping and working iron; these, in turn, were passed on to other branches of metalworking, such as the manufacture of locks and fixtures. Moreover, during Jones's tenure, Tula built several machines for Izhevsk and Sestroretsk. "Thus Tula," asserts Britkin, "gradually became one of the centers of a growing Russian machine building industry and a source of improved methods of arms production."[80]

A. A. Aleksandrov claims that the technology of the Ural armory was borrowed from Tula. For example, the installation of a double-barrel boring machine allowed a semiskilled operative to bore up to twenty-four barrels in a shift. However, other machines were under-utilized. A stock-making machine was available, but, Aleksandrov argues, armory authorities favored cheaper hand labor. In stock making, lock stamping, and barrel forging, hand labor prevailed until 1870.[81]

Sestroretsk, too, established specialized production of tools and gauges during the first half of the nineteenth century. In order to introduce greater uniformity in handmade percussion muskets, a model workshop was opened to make molds for measuring tools. Fifteen of the most skilled gunmakers were transferred and the shop opened in 1851. Although artillery officers in theory ran the shop, in practice the armorers themselves organized the technical work. The purpose of the work at the model shop was not only to attain uniformity of parts but also to provide a "school" whose "graduates" would spread their skills and production practices. Unfortunately, according to a history of the Sestroretsk armory, the laudatory goals of the workshop were thwarted. The workshop never did achieve uniformity and was costly to operate. Moreover, the factory administration allegedly was skeptical of innovation; thus whatever advances were made at the workshop were not integrated into the armory's production practices.[82]

It would appear that by the 1820s and the 1830s, the Tula factory had special-purpose machines that compared well with the metal-working technology then available. At this point, a critical look at the sources reporting on gunmaking practices is appropriate. In all likelihood Gamel' was not an impartial observer. Commissioned by the imperial government and dedicated to Nicholas I, no doubt his study was calculated to tell the new emperor what he wanted to hear. Jones, too, may have indulged in this. When Jones told Nicholas I that some of the Tula gunmakers were "*now* so skilled at making tools and machines that you would not find any better in England," he, too, clearly said what the emperor wanted to hear. In this way Jones praised not only the Tula gunmakers but also their British supervisor— himself. Indeed, Nicholas I "expressed his royal good will to Mr. Jones in the warmest words."[83]

Furthermore, Gamel' was not a specialist in machinery and may not have had firsthand experiences of armory practices at home or abroad and thereby may have exaggerated the innovations at Tula. Although it is impossible to confirm by internal sources, it is quite likely that Gamel', like the tsar, was given a showcase tour and was shown model workshops, new machines, orderly work processes, and diligent workers. The assembly of muskets from scrambled parts was a common demonstration, but it was difficult to ascertain whether the parts had indeed been selected at random.[84] The Tula gunmakers had quite a reputation for showmanship. According to one account, the gunmakers enjoyed making cute objects (such as shoes for the toy fleas in Leskov's story) and performing tricks for visitors. For example, a gunmaker would deliberately crack a barrel during fitting in a way that the untrained eye would not notice. Then the gunmaker would take the barrel and effortlessly break it in two over his knees. One source questions whether the magical assembly of a lock from random parts was as useful as the production of arms that fired surely.[85]

The most important question was not whether Jones actually did innovate, whether the Tula craftsmen were in fact clever, or whether the account by Gamel' was company hyperbole. Even if we assume that certain Tula shops could do all that was claimed, Russia failed to domesticate and to reproduce a technology it at one time had. Why did Russia need to borrow technology abroad a few decades later, and why by mid-century did domestic machine technology fail to reproduce itself and keep pace with foreign innovations? A full exploration of this problem must be deferred to the final chapter. However, here a few preliminary remarks may suggest some answers and, at the same time, conclude the analysis of Russian armory practice preceding the era of the Great Reforms.

A partial explanation is that the machine technology possessed by the Tula armory was not wholly domestic in the first place. The startling account of the wonders introduced by Jones presents a dilemma for Soviet historians. Britkin, for example, credits native innovators such as Zakhava at Tula and Deriabin at Izhevsk with anticipating many of Jones's improvements. He cites Gamel' to support his claim that significant innovations were developed at Tula during the years between 1818 and 1826 but neglects to mention that Gamel' attributed all improvements to Jones. Moreover, seeking affirmation of the high level of skills of the Tula gunmakers, Britkin cites Jones's high opinion of them, communicated to Nicholas I. He neglects to add, however, that Jones considered the gunmakers assigned to him initially inadequate.[86] The point is not to affirm or to deny instances of individual genius but to account for the broad diffusion of innovation and of new techniques.

Even if one credits individual genius and gives native Russian craftsmen their due, tsarist and Soviet histories are silent regarding the decades after Jones's, or any other mechanic's, innovations. Reading the various armory histories, one wonders what happened? Tula's Soviet historian, V. N. Ashurkov, for example, observes that after the Crimean War, European armies were provided with new and improved weapons, while "the backwardness of tsarist Russia held back the rearmament of the Russian army."[87] Prerevolutionary histories, while less accusatory, still blithely state that *beginning* in 1872 the stamping and forging of lock parts was done exclusively by machines.[88] One wonders what happened to Jones's "superb" die-forging machines or, for that matter, the die-forging machines allegedly at Sestroretsk before 1800?

The claim by Gamel' that all gun parts were made by machine does not necessarily mean that the entire armory was mechanized. Zagorskii admits that eighteenth-century boring machines "did not much outlive Peter the Great. Soon after his death . . . these machines were abandoned, fell into disrepair, and ceased to exist much before the year 1738; the Tula arms makers returned to manual production of all parts of weapons."[89] Britkin, otherwise a defender of native genius, states that the first steam engine, brought from the domestic St. Petersburg firm of Berd in 1811, sat idle for thirty years: "In actual fact, nobody ever succeeded in starting the engine and for a long time water remained the main source of power at the factory."[90] Many operations were still done by hand, as was true everywhere. This particularly applies to the final filing, fitting, and assembly stages, the most time-consuming, complex, and expensive operations. Frequent references by Gamel' to nearly uniform parts suggest that manual filing and finishing were not obsolete operations. Similar ambiguities mark the history of production at Sestroretsk. It is not clear either from the 1798 order of the War College cited earlier or from accounts by Soviet historians whether machinery was developed to achieve interchangeability. The problem is complicated by the fact that stamping machines installed in 1810 "permitted uniform gun parts . . . especially in the finishing of locks."[91] If the installation of stamping machines in 1810 permitted a high degree of uniformity, then what machine had been used to achieve interchangeability twelve years earlier? The terms *uniformity* and *interchangeability* were relative and changed over time; what one inventor or one Ministry of War circular called interchangeable one year may not have satisfied future standards of the same term.

Continued purchases of machinery abroad suggests important disincentives to mechanization and an inability to reproduce machinery and to integrate it into production processes. Zagorskii identifies three important disincentives at Tula. First, the gunmakers paid the armory for the use of machines. "This payment was so high that the arms makers derived no profit from the use of machines. . . . The mechanization of the work did not increase but decreased the earnings of the arms makers and they naturally resisted this mechanization."[92] Second, work on the machines had to be performed at the factory. Work at private workshops, "where officers rarely make their appearance," was less supervised; its rhythms could also include outside jobs. Third, the state did not reward invention. According to the armory administration's testimony before a military commission in 1755, cited by Zagorskii,

> No awards are given from the state treasury for inventions by the expert arms makers at the Tula plant because each expert is paid piecewise and does ordinary weapon making work for the state. And if anyone decides in his own interest and for his own profits to invent something new he does it with his own money and this the plant office does not prohibit.[93]

Early on, such disincentives to innovation resulted in an inability to reproduce machinery and to integrate it into production processes. According to Zagorskii, "Toward the end of the 18th century arms makers again began making weapons at home and drilling machines and machines of other kinds were left to be rediscovered. The technological processes in weapons manufacture at the end of the 18th century did not much differ from those at the beginning of the century."[94] To put it differently, and more familiarly, they had to reinvent the wheel. Likewise, Jones's machines were impressive but did not affect the organization of production or overall mechanization. Of 122 operations described by Gamel', only 28 were mechanized.[95] As one witness before the British Parliamentary Select Committee on Small Arms put it, "As far back as 1830 it struck me, when I heard how it was with the Russians, having their machinery as they do, and yet that they could not manufacture them cheaper by machinery; and I heard that their machinery was very complete; but it does not appear that they can manufacture them by machinery if they come to Belgium for them."[96] As Prosser suggested, the Russian machinery was by 1854 "too complicated" and English machinery had improved much in twenty years.[97] As we have already seen, full mechanization was never a high priority in the various proposals to reorganize the armory in the Nicholaevan era; the state of the art in 1817 was too difficult to integrate into production processes.

Conclusion

The gunmakers' closed corporate structure, perpetuated generation after generation, coupled with the absence of a large private arms market, minimized contact with the outside world. Outsiders, who frequently bring innovations, were responsible for new methods at Tula. In the eighteenth and early nineteenth centuries, a few British and German workmen introduced European arms-making technology. The most noteworthy of these individuals, of course, was Jones. Even native mechanics, such as Zakhava, came to Tula. But such examples of influential and innovative outsiders are rare.

Although the armories produced for an "outside" consumer, this consumer was the government, which showed only passing interest in changing armory practice, particularly at the largest armory, Tula. There were never enough gunmakers in the private sector whose numbers could complement the government armorers. Nor did there need to be. The surplus gunmakers indentured to the armories but engaging in outside work were available to fill government orders. Likewise, the large number of semiskilled and unskilled laborers indentured to the armory enabled common labor to be commandeered from within rather than purchased from without.

The shop system continued to have considerable impact on the nature of work. The component parts of firearms were contracted out, and most of the work on parts was done by hand at home. Only a small portion of the work was actually done at the small arms factory. More competitive and skilled, the Tula armorers presumably were in a better position to innovate. Yet, as we have seen, they worked largely at their own small workshops. Thus, even had the Tula gunmakers been innovative, the Tula armory would not have reaped the benefits.

Even though Sestroretsk, and to a lesser extent Izhevsk, centralized work in one place, the factory system and the attendant advantages in supervision were merely grafted onto the preexisting shop organization of work. Since neither the work nor the armorers could be supervised, it was difficult for the chief engineer to enforce quality control of the work performed in numerous workshops. Although the foreman had authority over the shop, it is not clear whether he was always able to exercise that authority effectively. According to the management at Izhevsk, "You have to be an extraordinarily energetic, fair, and independent foreman in order to stand up to the rapaciousness of the stewards and other petty authorities."[98] Perhaps most disadvantageous in the long run, dispersed gunmakers were less able to train each other and to learn from each other. This factor acted as a further brake on the diffusion of technology within the industry.[99]

Indentured to the armories, skilled gunmakers and other laborers neither became independent artisans nor developed craft pride. P. Glebov, commenting on an official report of the Tula armory, concluded that gunmakers were in poor straits because they were virtually serfs of the factory administration. Writing after the emancipation, a mining engineer claimed that "indentured labor, in the melancholic admission of engineers of that time, inflicted a kind of apathy on everyone."[100] Many nineteenth-century studies of the Russian gun trade suggest that by and large meticulous workmanship and even craft pride were wanting among the gunmakers. Russian gunmakers made weapons for the army, not for commerical sale, and the Russian gun trade demanded few custom-made guns. The best-made gun was rarely regarded as a work of art, as it occasionally was at Liège and Birmingham. The small private market in Russia for firearms provided weak stimulus for the development of a large number of skilled gunmakers and for the development of craft pride. When foreign gunmakers went from custom-made to machine-made products, argues the history of the Sestroretsk armory, they retained craft pride. Moreover, Liège craftsmen, and allegedly all foreign craftsmen, were used to machines, "a part of life" not commonplace to Russian gunmakers.[101]

Yet Russian armory practice was ideally suited to Imperial Russia's needs during the first half of the nineteenth century. The closed corporation reflected Russia's prevailing corporate structures and the obligations and immunities of service. The shop organization preserved the available craft skills, especially in Tula. The subcontracting system and work at home saved the state money, gave the gunmakers at least part-time work, and prevented the loss of skills. The government armories and planned procurement adequately supplied a peacetime military. The number of gunmakers was not large for Europe's most populous nation; nevertheless, there were enough gunmakers to fill state orders during a period of modest increases in weapon procurement. The periodic purchases of arms and machines abroad, coupled with gradual native improvements, was adequate during a time of gradual changes in weapons systems and in metalworking technology. However, Russia's military needs in an era of reform and rapidly changing weapons design marked a sharp break with the past. New challenges suddenly emerged in the adoption of small arms, in the sources of supply, and in the role of firepower in tactical doctrine.

AMERICA AND THE RUSSIAN RIFLE

After all the efforts of the Peace Congress, it is now universally admitted that the time has not arrived when we may turn our spears into pruning-hooks. On the contrary, there never was a period when men were so bent upon applying them to a deadly use, rendering them still more destructive, and giving them the keenest edge that steel will take. . . . Who would now go into battle with Cromwell's matchlock, or lumbered with Brown Bess? We might just as well take David's sling. The armourer's art, indeed, is making such strides—a year, a month, a day gives it such novel developments—that we are really puzzled to know what weapon is the best. How shall we arm?—with carbine, rifle, or revolver? If with revolver or rifle, or both, with which and whose?

—"Firearms," Colburn's United Service Magazine
(September 1859)

● THUS AN ENGLISH MAGAZINE SUMMARIZED the craze of rearmament that engulfed Europe after the Crimean War. The period between the Crimean War and the Franco-Prussian War saw feverish activity to rearm infantries and cavalries with rapid-fire small arms, manufactured in mass quantities. This period saw the greatest amount of military activity since the Napoleonic Wars, and the American Civil War and the Prussian Wars provided the first large-scale testing grounds for the new firearm technology. Percussion ignition had replaced the flintlock; the Minié and similar conical balls had been introduced but soon had been superseded by the metallic cartridge containing powder, ball, and fulminate; the rifled bore had replaced the smoothbore; steel had replaced iron for the barrels, which were made with a smaller caliber; breechloaders had replaced muzzleloaders; and, finally, carbines and revolvers had been widely introduced for mounted troops. As a result, new weapons systems featured reduced reloading time and number of misfires, flatter trajectory, longer range, greater velocity, and greater accuracy.[1]

One year before *Colburn's United Service Magazine* in England noted the feverish pace of rearmament, the lead article in a military journal that was just beginning publication on the other side of Europe also noted the improvements in small arms attracting everyone's attention:

> The improvements made in the past decade in small arms and applied during the last war could not help but attract much attention. All nations are now taking the most energetic measures to equip the infantry insofar as possible with the latest firearms, undertaking tests, and forming committees to discuss the armament and organization of the army. Papers and journal articles are appearing everywhere which examine from various points of view the influence of rifled arms on military action. Nowhere, however, is there such passion in opinions, such an irritated tone in references to the past, as in Russia. . . . In all articles published on this subject, and especially in conversation, there has been no mercy shown to our previous regulations or to our rules of tactics.[2]

Russians had good reason to express "such an irritated tone in references to the past": Russia's defeat in the Crimean War was partly attributable to its backwardness in arms. The editors of *Voennyi Sbornik* were quick to point out that at the outbreak of the war the problem of the best weapons system had not been solved even by the victors; that no one was advocating equipping the entire infantry with rifles; that to have rearmed Russia during the conflict would have required one million arms and would have been prohibitively costly; and that the glory of the Russian soldier was even greater in the face of the enemy's superior weapons. Nevertheless, it was clear that such assurances could no longer justify inactivity.[3] As the Artillery Department stated to the Ministry of War, "Russia cannot, and indeed, must not lag behind the other powers in the sweeping rearmament of its army whatever the cost to the state."[4]

Minister of War Dmitrii Miliutin, in his report of 1862 outlining plans for sweeping military reforms, starkly stated the lessons of Crimea:

> In the present state of the art of war, technology has become extremely important. Improvements in weapons now give a decided advantage to the most advanced army. We became convinced of this truth by the bitter experience of the last war. Because they were poorly equipped with rifled arms, hastily converted from smoothbores, our soldiers had to endure heavy casualties. . . . This war has made us recognize the necessity of the most energetic

measures to equip our soldiers with up-to-date arms. . . . Now we must honestly admit that we are materially behind other European countries in our munitions and weapons.[5]

The 150-year-old relationship between army and society was shaken by the Crimean War. The Great Reforms launched by the new emperor, Alexander II, and by enlightened bureaucrats such as Miliutin freed the serfs and introduced universal military service. Laborer and soldier both were now free. The laborer was no longer indentured to the government arms factories. The virtual lifetime duty of the soldier was replaced by six years of active duty and nine years of reserve duty. A greater recognition of firepower gradually permeated tactics and training. The Ministry of War undertook "the most energetic measures" to equip the infantry with modern weapons, and in the process the state became more directly involved in the development of the arms industry than ever before.

The Beginnings of Rearmament

Rearmament implied adoption of new weapons systems. In an effort to disseminate information about military affairs, engineering, and developments in small arms, the government sponsored several new publications. *Oruzheinyi Sbornik* (Small arms review), a quarterly founded and edited by V. Bestuzhev-Riumin and V. Chebyshev, was the first private military periodical in Russia. This journal specialized in small arms technology—the theory and history of firearms, the description of different systems in Russian and foreign armies, and the state of small arms factories in Russia and abroad. Describing the goals of *Oruzheinyi Sbornik* before an audience at the guard artillery in 1861, Chebyshev claimed that small arms had not been considered especially important, particularly among infantry officers. Although small arms had aroused a flurry of interest at the end of the Crimean War, Chebyshev was concerned that apathy might again set in. Thus the goal of the journal was to familiarize the public with current developments in firearms and, even more important, to give the pubic an opportunity to keep abreast of important military issues of the day.[6]

With or without an informed public, Russia still had an autocratic government exercising its power through ministries or their equivalent. Minister of War Miliutin created a series of committees to collect technical descriptions, to supervise tests, to evaluate the suitability of new arms, and to recommend adoption of new systems. The first of these committees, the Committee on Improvements in Small Arms, created in 1856, considered the adoption of new muskets at the end of the Crimean War. Five years later this committee was succeeded by the Commission of Small Arms. In 1869, at the beginning of the most intensive period of small arms development, the ministry created two temporary commissions to supervise the adoption of breechloaders and metallic cartridges. These commissions were the Executive Commission of Rearming the Army of the Artillery Department, chaired by Lt. General Rezvyi, and the Main Executive Commission for Rearming the Army, chaired by Miliutin himself.[7]

Russian rearmament after the Crimean War proceeded in two phases. In the decade up to 1866, the rifled bore replaced the smoothbore; after 1866 the breechloading rifle replaced the muzzleloader. Both phases also included many other changes, and at one time or another virtually every system in use in Europe and America was considered for adoption. Moreover, the practice of introducing newer systems on a selective basis meant that at any one time different military units were armed with different systems. For example, in 1863, when all infantry battalions were scheduled to receive new .60 caliber rifles, the standard arm of the dragoons remained the .70 caliber percussion musket used in the Crimean War, while the standard arm of the regular cavalry remained the 1818 flintlock musket.[8] This inconsistency in the process of rearmament was a bane to Russian performance during the Russo-Turkish War of 1877–1878.

Russia was defeated in Crimea by arms possessing greater accuracy and longer range. Russia's infantry arms were inferior not in their firing mechanism (a muzzleloading percussion musket was the standard arm for both sides) but in their barrels, virtually all of which were smoothbores. After the Crimean War, the Russian army attached high priority to accuracy and range in its selection of system design. Reducing the caliber without rifling, or rifling the bore and using the conical bullet improved accuracy and increased the range of fire. Like other governments implementing low-cost rearmament, the Russian government decided to produce rifles initially by rifling surplus smoothbores. Contracts for conversion models were concluded with Colt and with the Belgian manufacturers Fallis and Trapman. The latter manufacturers agreed to rifle one hundred thousand new barrels over a period of two and one-half years in special shops established for this purpose at the Izhevsk and Tula factories, to supply rifling and boring machines and tools, to loan twenty-four master workmen, and to train Russian workers.[9]

The government also began to adopt newly made rifles of both domestic and foreign manufacture. In 1857, a .60 caliber muzzleloading infantry rifle was adopted; the first thirty thousand Russian-made rifles were equipped for the sharpshooter battalions. Beginning in 1859, the Cossacks were equipped with a .60 caliber rifle made in Liège.[10] Because the Belgian and Russian .60 caliber arm was a newly made rifle, not a converted smoothbore, it was called a "rifle" (*vintovka*) rather than a "rifled arm" (*nareznoe ruzh'e*). The Small Arms Division of the Artillery Department assigned more than semantic significance to this nomenclature:

> This name "rifle" will more likely convince the soldier of the worth of his arm; by comparison, the name "rifled arm" suggests to the soldier that it is a simple arm with only certain changes which he does not completely understand.[11]

The .60 caliber rifle was still a muzzleloader, dangerously slow and cumbersome to load. After the American Civil War and the Prussian successes against Schleswig-Holstein and Austria, the importance of rapid fire could no longer be neglected. All over Europe tests feverishly began on breechloading rifles. In the mid-1860s, European armies moved quickly from tests to adoption, and, while the Russian army still had muzzleloaders, the major European powers had adopted breechloaders. Russia had no choice but to follow. The pace of change was dizzying, prompting Miliutin to write: "We had no sooner finished rearming our army with .60 caliber rifles, than the question of breech-loaders using metallic cartridges came up. This was so new, that every-one was testing the first, and very imperfect, inventions, which were suddenly forgotten with the appearance of new mechanisms."[12]

Even though many inventions and improvements could be tested ad infinitum, as Miliutin suggested, some system had to be adopted, manufactured (or purchased), and provided to the infantry and cavalry. Since making new breechloaders from scratch was prohibitively expen-sive, particularly at a time of rapidly changing system designs, most governments tried to find the best lock mechanism to convert muzzle-loaders into breechloaders.[13]

In 1866 the Ministry of War ordered the small arms factories to convert 115,000 .60 caliber muzzleloaders into breechloaders using a percussion conversion system designed by the English gunmaker Terry and improved by the Tula engineer I. G. Norman. However, equipped with a capsule containing a paper cartridge with powder and bullet, the Terry-Norman rifle did not provide the rapidity of fire avail-able in the needle systems and was considered inferior to European models.[14] At the same time, the English gunmaker Carl offered the Russian government his modified needle system using a sliding bolt action and fixed round paper cartridge. The small arms commission had earlier rejected the Prussian needle gun, despite its rapid fire, on the grounds that its lock mechanism was excessively complex, the spiral spring that activated the needle was easily damaged, and the needle itself was too delicate and potentially difficult to replace, especially on the battlefield. However, in its simplicity of construction, durability, and rapidity of fire, the Carl rifle claimed to be an improvement over the Prussian needle gun and over the Terry-Norman. Perhaps most important, the Carl was easier to convert than the Terry-Norman. By mid-1867, conversion of muzzleloaders by the Terry-Norman system ceased after only half of the original order had been filled, and two hundred thousand muzzleloaders were ordered to be converted by the Carl system at the small arms factories.[15]

The same year, two other conversion systems, one designed by Lieutenant N. M. Baranov, director of the Naval Museum in St. Petersburg, and the other designed by the Bohemian baron and gunmaker Sylvester Krnka were submitted to the small arms commission. A variant of the English Snider, the Krnka was a breechblock rifle using a metallic cartridge. It had been rejected in Austria but was admired in France and Russia. Like other conversion systems of the period, it was fragile; but it was considered superior to the Carl. Moreover, although the Baranov system was regarded as better militarily, the Rezvyi commission was persuaded that the Krnka could be manufactured more quickly and cheaply than the Baranov model, later adopted by the navy. Although the conversion was more difficult than the Carl conversion, in 1869, Russian small arms factories were ordered to convert 183,000 smoothbore muskets and make 120,000 new rifles.[16]

Russian Colonel Among the Yankee Arms Makers

At the same time the Russian army was hurrying to convert smoothbores into rifles and muzzleloaders into breechloaders, reports were arriving from the United States about recent improvements in small arms. An English correspondent from Richmond, quoted in the Russian *Inzhenernyi Zhurnal* (Engineering Journal), stated, "All experts confirm that there has never been as instructive a war for artillery officers and engineers as the American War. Never before have so many interesting inventions been applied in military science and never before have such positive results been obtained."[17]

The Civil War brought about a rapid adoption of small caliber breechloading arms using metallic cartridges. European officers on American soil during the Civil War, including Russian naval officers, brought back samples of new system designs, and small arms commissions all over the world were testing Peabody, Remington, Ladley, Morgenstern, Winchester, Jenkins, Green, Berdan, and a host of other firearms systems. As one Russian officer wrote in 1867, "It is difficult to disagree on the practicality of the North Americans, especially when it comes to firearms. They have clearly shown to the whole world that they are number one in firearms."[18] Moreover, the United States had the most advanced methods of small arms production and, as one Russian artillery officer put it, "its famous metal factories hummed with arms orders during the protracted Civil War." However, with the rapid drop in the domestic military market at the close of the war, American gunmakers desperately sought foreign sales.[19]

Given the reservations of the small arms commission about the Prussian needle mechanism, the Russians were quick to recognize the potential of the American metallic cartridge. Indeed, in Russia, the rim-fire metallic cartridge was known as the "American cartridge."[20] Before the American Civil War ended, the Russian government sent Colonel of Artillery Alexander Gorlov (1830–1905), a small arms expert, to the United States to study American rifles, to follow tests on weapons, and to gather detailed information on the use of metallic cartridges. Before he finished his work among the arms makers of the Connecticut Valley, Gorlov almost singlehandedly selected weapons systems, modified American designs to meet Russian specifications, and organized and supervised production of Russian orders in American factories. As a result, a War Ministry that had proceeded very cautiously and somewhat erratically in the choice of conversion models virtually overnight adopted modern rifles, cartridges, and revolvers.

Like the myriad of middle- and lower-level Russian officials, civilian and military, Alexander Gorlov has left barely a trace of evidence of his life. Born in 1830, Gorlov became an artillery engineer after graduating from the Mikhailovskii Artillery Academy. Beginning in 1851, Gorlov served as scientific secretary of the Technical Committee of the Main Artillery Administration. Gorlov began writing on firearms in the 1850s, and as early as 1859 *Artilleriiskii Zhurnal* printed his small "Anglo-Russian Lexicon" of 340 technical terms. He went on his first foreign assignment in 1859, perhaps to England. His article in *Morskoi Sbornik* extensively covered Colt and was one of the first to appear with information concerning revolvers. In 1866 *Oruzheinyi Sbornik* and *Artilleriiskii Zhurnal* carried his report on the use of breechloaders and metallic cartridges in the American army and his series of wartime sketches. According to Department of State records, in 1867 Gorlov was attached to the Russian legation in Washington; one year later he was registered as a military attaché residing in Hartford.[21]

Gorlov returned from his first American voyage with rifles and

machines to make metallic cartridges. As the small arms inspector V.
Buniakovskii later described it, the issue of Russian adoption of metallic cartridges was for all intents and purposes settled quickly when
Gorlov returned from the United States with his booty. His 1866 report on the use of rifles and metallic cartridges in the U.S. army convinced the Artillery Department of the advantages in firepower and
accuracy of small caliber arms and the advantages in rapidity of fire
of breechloaders employing metallic cartridges. Although the guns and
the cartridges were in an early state of development and the small
arms commission did find deficiencies in the cartridges themselves,
early tests demonstrated that several American systems had superseded the needle gun.[22]

In 1867, Gorlov, this time accompanied by Captain K. U. Gunius,
a member of the Technical Committee of the Artillery Department
and secretary of the small arms committee, returned to the United
States to find a suitable metallic cartridge for adoption in Russia. Ed-
uard Stoeckl, minister to Washington, informed Colt's vice-president
and general agent, William Franklin, that Gorlov was going to the
United States with "an important commission. My government in-
tends to order in the United States fifty or one hundred thousand new
muskets." Stoeckl advised Gorlov "to go straight to Franklin who is
an honest man and who is at the head of the largest manufactory
of arms in the United States."[23] When the Russian officers decided
to conduct their work in Hartford, Franklin shrewdly offered them
the facilities of Colt Patent Firearms, thus establishing a business
friendship that was to continue for almost two decades.[24]

Upon arrival, Gorlov and Gunius closely followed the progress of
a special commission of the U.S. army that was testing new rifle models for the most appropriate caliber, length and weight of barrel, weight
of bullet, cartridge, and powder, and amount of rifling. In addition, the
Russian officers attended many private and state tests. For example,
three days after their arrival in New York, they were invited to observe
extensive tests of twenty-five musket conversion systems for the New
York state militia. During the tests, many manufacturers and patent
holders presented to the Russian officers samples of their wares, hop-
ing to receive large orders from the Russian government. In this way,
Gorlov acquired many models to take back to Russia for additional
tests. Finally, the U.S. Patent Office gave Gorlov and Gunius open
access to its information on inventions and improvements. Gorlov
must have been overwhelmed with the opportunities available:

> The success of the Spencer carbine decisively showed the importance of hav-
> ing a good model of a rapid-fire arm. This circumstance, when combined

with the desire of successful inventors to make a big profit all at once, has made, as frequently happens in the U.S., feverish furor and excitement. Many people from all walks of life, who frequently have little understanding of firearms, and even a few personages of the female sex, have applied their effort and imagination to creating a model firearm for the army. It is unbeliev- able how the number of patents issued for such inventions has suddenly increased.[25]

There were many American rifle and cartridge models from which to choose. Arriving in the United States at a time of intensive activity on the center-fire metallic cartridge, Gorlov and Gunius designed rifle and cartridge models, undertook tests of their own, and ordered a small number of the best .45 caliber rifles, cartridges, and revolvers for final system selection.[26] From Colonel Gorlov's stay in the United States came three modernizations in Russian small arms: the Berdan infantry rifle, metallic cartridges, and the Smith and Wesson revolver.

The "Russian Rifle"

Despite the "furor and excitement" of all the inventions and im- provements, Gorlov and Gunius found none of the systems they tested to be perfect. Of the rifle systems available, the best combination of breechloading mechanism and metallic cartridge appeared to be that designed by Colonel Hiram Berdan (1823–1893), an American inventor of several improvements in small arms, particularly the "breechlock," a mechanism for closing the breech of a rifle, and the designer of a trap-door action used by the U.S. army after the Civil War to con- vert muzzleloading Springfields into .45 caliber breechloaders.[27]

Berdan's trap-door action presented several advantages to the Russian officers. To begin with, the previous breechloading sys- tems—the Terry-Norman, the Carl, and the Krnka—were all designs to convert old muzzleloaders. The Berdan rifle shared many similari- ties with the Russian muzzleloaders then in use, thus promising an easier and cheaper conversion to a breechloading arm.[28] Berdan was willing to make certain improvements in the lock mechanism and cartridge and adjust them to the .42 caliber bore desired by the Rus- sians; this provided a flatter trajectory, higher velocity, greater accu- racy, greater penetrating power of the bullet, and less weight. After testing many models, they settled on the Berdan model .45 caliber center-fire cartridge as a basis for a modified design by Captain Gu- nius, soon used in the Krnka conversion system.[29] Between 1867 and 1868, Gorlov and Gunius made twenty-five modifications, including a reduced caliber, the elimination of the external hammer and an improved bolt construction, and a new cartridge ejector mechanism. According to Captain Buniakovskii, later inspecting Russian orders at the Colt factory, Gorlov and Gunius effectively designed a new rifle.[30]

In 1868 the Russian government adopted the small-caliber Berdan infantry rifle and purchased full rights from Berdan to order abroad or to manufacture the rifle in Russia. Russia then contracted with the Colt Company for the manufacture of thirty thousand rifles and with the Union Metallic Cartridge Company of Bridgeport, Connecticut, for the manufacture of seven and one-half million center-fire metallic cartridges. Delivery to Russia commenced in early 1869.[31] In the end, the artillery office boasted in a report to Minister of War Miliutin, the Russian army would have an arm that would compete with the best. According to the authoritative *Small Arms of the World*, this rifle and cartridge marked "the beginning of the era of high-power small bore rifles with great range and accuracy and relatively flat trajectory."[32]

The contract provided an excellent business opportunity for the Colt Company. Having just rebuilt the factory after a disastrous fire in 1864, the company was eager to set up production for the Russian contract. According to William Jarvis, the Russian order, the first large foreign order after the fire, would "make business lively at the Armory for a good while." Perhaps for this reason, Colt was willing to get a lower rate of profit than customary to compensate for labor and overhead.[33] Buniakovskii predicted, "After completion of the Russian order, the Colt factory will make rifles of the Russian system for other countries. . . . In all probability, the Colt factory will receive an order from a European government, since so many military agents of foreign governments have appeared here."[34]

Colt eventually received the Russian order in 1868 due to a variety of factors, among which only one was mechanization of the Colt factory. The company's success at winning this contract was in part due to Samuel Colt's earlier machinery and musket sales to the Russian government. Certainly, the contact had already been made, and, although Samuel Colt's dealings with the Russians were not entirely successful, the fact remains that Gorlov was instructed to see the company's vice-president, William Franklin, upon his arrival in the United States. As Russian sources frequently claimed, Colt was eager to take advantage of the skills in system design offered by Gorlov and Gunius; moreover, the insistence of the Russian officers that Colt strictly fulfill all its obligations kept production moving and even resulted in a greater degree of interchangeability than originally expected.[35] Another factor working in Colt's favor was that William Franklin had already offered space in the factory to Gorlov and Gunius so that they could conduct their tests on rifle and cartridge systems. But perhaps the most important factor in Colt's success in getting the Russian contract was the concurrence of the Russians' needs in system testing and alteration and the manufacturing methods practiced at Colt.

The small arms committee was still not completely satisfied with the 1868 Berdan model (or, more precisely, the Berdan-Gorlov model). In 1869, Berdan visited St. Petersburg and presented a new rifle with a sliding-bolt action and with the improvements by Gorlov and Gunius already incorporated. The small arms committee was impressed and, after favorable initial tests, proposed adoption of the new rifle. In October 1869, the Russian government signed a licensing agreement with Berdan to purchase thirty thousand sliding-bolt action Berdan infantry rifles, dragoon rifles, and cavalry carbines from the Birmingham Small Arms Company. The improved Berdan was usually referred to as the "Berdan 2," "the 1870-model infantry rifle," or, colloquially, "the Berdanka."[36]

The Colt Company, of course, hoped that the Russian government would continue to order the 1868 model. In December 1869, William Franklin visited St. Petersburg in an attempt to get more orders for the 1868 model rifle. According to Franklin, the Ministry of War had decided to

adopt Berdan's new system, not because the old pattern is considered a bad gun, but because the new gun is heard to be better, being simpler in construction, easier to manufacture and more rapid in firing, requiring one movement less in loading. The great objection to the old gun is the difficulty of ejecting the cartridge shells.

"Our gun," Franklin added, "had no friends on the Small Arms Committee," which "rejected our gun long ago."[37] Despite Colt's initial success in getting orders from the Russian government for fabrication of the Berdan 1, competitors, in particular the Birmingham Small Arms Company and the Leeds manufacturers Greenwood and Batley, also sought contracts for the Berdan 2. As an inducement to the Russians, the Birmingham Small Arms Company made the unusual offer to sell all the machinery and tools used in making the rifles to the Russians at the end of the contract.[38] Berdan, lobbying in St. Petersburg during the autumn of 1870 on behalf of his new system, likewise was authorized by Colt to offer "to sell all the machines they now have that they use on this contract when the contract is finished."[39] Such negotiations reveal Russia's long-range intentions. Although Franklin believed that the government "might entertain a proposal from us for [30,000] if the price were very low, . . . their notion is to manufac-

ture hereafter in Russia." Colt's London agent had learned from contacts that the Russian government "had resolved in the future to make
all their arms and cartridges at home."[40]

Praise for the rifle came from many quarters. In the opinion of Captain V. Buniakovskii, a small arms inspector for Russian orders at the
Colt factory, the strength of the mechanism, the ballistics qualities,
the ease of replacing parts, the accuracy, and the durability made the
Berdan rifle an outstanding military arm. The author of a Russian
textbook on firearms, V. Shkliarevich, stated that "the accuracy, ease
of manipulation, lightness, sturdiness, and the simplicity of construction which allows assembly and disassembly without a screwdriver
all make the Berdan model an excellent military arm."[41] Captain N.
Litvinov, later important in the campaign to establish a protectorate
over the Central Asian khanate of Khiva, echoed Shkliarevich:

> I must describe the favorable impression this arm made on the battalion
> and the absolute confidence in the fine qualities of this weapon among both
> officers and soldiers. All realize that the units armed with the Berdan rifle,
> which our government has successfully acquired, will be almost the best
> army units in Europe today.[42]

Minister of War Miliutin stated that the Berdan "may be recognized
as just about the best of all existing firearms."[43] After becoming the
standard infantry rifle in 1868, the Berdan was proved durable in Turkestan in 1870, in the Russo-Turkish War, in the Afghan campaigns,
and, although it was later replaced by the Mosin as the line weapon,
even in World War I, in the Russian Revolution, and in the Civil War.[44]

Russian officers also praised the Berdan for another feature: the so-
called Berdan 1 was essentially a Russian rifle. In America the Berdan
1 was known as the "Russian rifle," although few knew the names
of the Russian officers who had designed the weapon. The standard
survey of American firearms systems and of the technologies of their
fabrication at the time stated that accounts of trials of different rifles
frequently cited the names "Berdan (Russian)" or "Berdan (Russian
model)" and that the rifle "has never borne any other designation than
the name of its inventor, though designed and manufactured in the
United States and Europe, under the supervision of a Russian officer,
selected for his mechanical proficiency, General Gorloff."[45] Foreign origin often lends mystique to a product, and thus the name "Russian
rifle" attached a certain aura in America, the homeland of the inventor, just as the name "Berdanka" attached a certain aura to the rifle
in Russia, the homeland of the designer of the modifications.[46] Americans recognized the authorship of the rifle and, even more pleasing

to Russian officers, recognized is quality. Hartford's best marksmen, Gorlov noted, had won first prizes at shooting contests with their "Russian rifle," although elsewhere in America Gorlov believed that his rifle was not given a fair hearing.[47] Captain Buniakovskii, the inspector of orders at Colt, proclaimed,

> I consider it my duty to report, as an eyewitness, that in America they consider the Russian rifle and cartridge (the one we call the Berdan) to be the best; they also consider the Colt factory, which makes our sharpshooter rifle, and the Bridgeport factory, which makes our small-caliber cartridges, to be the best factories.[48]

Metallic Cartridges

Russian rearmament after the Crimean War, and the intensive search for new weapons systems in the late 1860s and the 1870s, coincided with the revolution in firepower brought by metallic cartridges. The high-quality American cartridges and the facility of machine-made fabrication were major selling points of the American arms industry to Russian small arms experts, especially to Colonel Gorlov.

Russian research on the military use of metallic cartridges began in 1864. Gorlov's visit to the United States introduced Russian officers to the metallic cartridge, and in particular to the Berdan model. In the mid-1860s the Russian Army had adopted several conversion model breechloaders. When the army decided to adopt metallic cartridges for the conversion models, it realized that it would need more than one hundred million annually. Since there was virtually no domestic production, military needs could be satisfied only by foreign orders. The Russian government placed frequent orders with the Union Metallic Cartridge Company (UMC) in Bridgeport, Connecticut, for the Berdan-Gorlov-Gunius .42 caliber center-fire metallic cartridges. Gorlov himself was sent to UMC to supervise manufacture of the initial order of seven and one-half million cartridges. According to an American expert on firearms and machinery at the time, "General Gorloff of the Russian army, who is well known for his knowledge of the subject of fixed ammunition, paid special attention, while in this country, to the manufacture of the Berdan cartridge."[49]

From Russian and American accounts it would seem that Gorlov gave new meaning to the concept of "paying special attention." In the design, production, and inspection of the Berdan rifle and the metallic cartridges, Gorlov earned a reputation as an exacting taskmaster. Inspecting finished rifles at Hartford in the summer of 1869, Gorlov complained to Colt Vice President William Franklin that workmen were filing a seam in the barrels to conceal a defect in the Firth steel. Gorlov charged that this "illegal work" was "obviously done with the view to deceive our inspection." The defect could cause the barrel to burst in the hands of the soldier firing the weapon, "causing the soldier's death and undermining entirely the confidence of our troops for the breechloaders made in your factory and in the United States in general." Gorlov added that Franklin already knew of the alleged "distrust in Europe for the goods manufactured in America, especially the arms." The Russian arms inspector Buniakovskii boasted, "I must also say that the engineers of both factories [Colt and UMC] privately admit that they owe the precision of their work to the Russian officers."[50]

American accounts agree. According to Alden Hatch, historian of the Remington Arms Company that later bought UMC, Marcellus Hartley, the founder and president of UMC, gave "General Gorloff part of the credit for the excellence of UMC ammunition. The almost impossible degree of perfection demanded by the Russian set a standard that the company always afterward maintained, a standard that exceeded reasonable expectations." Hatch gives a sensationalistic clue, though it cannot be verified, to Gorlov's demand for an "almost impossible degree of perfection." Beset by anxiety over his own performance, Gorlov allegedly only relaxed two years after arrival in the United States. "Do you realize," he reportedly confided, "that this is the first evening of pleasure I have had since I came to America?" Explaining the cause of his anxiety, Gorlov stated, "I am personally responsible to His Imperial Majesty for the success of this contract, for every cartridge you send to Russia. If this contract fails, the best thing I can do is to take one of your excellent American revolvers and blow out my brains."[51]

Gorlov's stay at UMC was crowned by the visit of Grand Duke Aleksei Aleksandrovich on 25 November 1871. Gorlov himself gave the Grand Duke a personal tour of the plant. Everyone dressed formally for the occasion, and every machine was decorated with flowers. The Grand Duke, a ladies' man himself, was astounded that the young women tending the machines wore silk dresses.[52]

The Russians were satisfied with the more than two million metallic cartridges they received from UMC. The cartridges had an almost miraculous longevity; according to one account, UMC cartridges that had been waterlogged for weeks after a ship transporting them to Russia ran aground in a gale fired perfectly. The metallic cartridges were deemed so perfectly suited to the Berdan conversion model, and gave such superior results, that both rifle and cartridge "managed to inspire an uncustomary trust" in soldiers.[53]

However, the Russian colonel had another mission among the Yankee arms makers. The Russian government was already planning domestic production of metallic cartridges. Gorlov inspected machines for making metallic cartridges and requested a set of presses, cutters, and millers at the Frankfort Arsenal from Secretary of War Stanton. As Gorlov reported to his superiors in Russia,

> Mr. Stanton immediately consented to my request and, in the name of the American Republic, presented me with a full series of the desired machines for transmittal to the Russian Government, as a symbol of the sincere friendship the one country bears to the other. At the same time, Mr. Stanton declined my offer to recompense the American War Office for the cost of the machines.[54]

Smith and Wesson Revolvers

"The shooting didn't last long. Who knows what they saw there, so full of smoke was the street. In my jacket I had my 'Smith and Wesson.' I felt safe.[55]

After Samuel Colt's visits to Russia, revolver experiments became more frequent, not only because of the proliferation of foreign and domestic models but also because the question of rearmament of the cavalry was frequently discussed in connection with the rearmament of the infantry. However, cavalry rearmament did not have the priority accorded infantry rearmament, and as late as 1872 an English observer wrote:

> In point of armament the Russian cavalry is somewhat behind the age, although, as in other branches of service speedy reform is promised. A curved saber, a pistol issued in 1839, which no one would dare fire, and a lance for the front rank men of all regiments, hussars and cuirassiers included . . . complete the weapons possessed by Russian cavalry men. A Berdan carbine and a revolver for each man are, however, spoken of.[56]

Although the number of weapons involved was smaller, choosing among available revolver systems actually presented more difficulties than choosing among infantry rifle systems. First, revolvers were made primarily by private manufacturers. While this did not hinder revolver development in the United States, with its ample number of private arms makers, it presented a serious obstacle in Russia, a country with a poorly developed private firearms market. Second, revolver tests were less standardized and therefore less reliable than those applied to infantry rifles.

Had the cavalry been the only consumer of revolvers, the difficulties might indeed have greatly delayed adoption of modern revolver systems. But the Russian government had an additional incentive for proceeding expeditiously. Frustration over the compromises and shortcomings of the Great Reforms promulgated by Alexander II mobilized a small but influential number of radical Russian youth. Mysterious fires in St. Petersburg, threats of violence, peasant riots, unrest in Poland, and, finally, an assassination attempt against the tsar in 1866 lent a sense of urgency to the acquisition of modern arms for Russia's infamous gendarmes. Responding to a suggestion from the chief of gendarmes that an alternative be found to the Lefaucheux revolver, the small arms commission stated, "In view of the urgency of arming our cavalry with revolvers, the officers sent to America have been ordered to follow carefully the improvements and simplifications that can be introduced into such weapons and accordingly to determine also the best revolver model for the gendarmes."[57]

Gorlov's mission to the United States presented the small arms commission with an excellent opportunity. Although Gorlov spent much time at Colt's Hartford plant, where the Berdan rifle was redesigned and manufactured, it does not appear that he paid much attention to Colt revolvers, which did not sell well in Europe in the 1860s. The Smith and Wesson Company, however, had been selling the revolver in St. Petersburg during the 1860s through several agents. Mention of Russian tests of the Smith and Wesson revolver can occasionally be found in military sources. Although the small arms commission found it the best cavalry arm for simplicity of construction, ease of fire, strength, and accuracy, the requirements for revolver adoption were so specific that no system, including the Smith and Wesson, was completely satisfactory. Later tests seemed to give preference to a Remington model pistol, which the Russian government was on the verge of adopting in 1870. (According to Hiram Berdan, in St. Petersburg to promote his rifles and also a pistol, many Russian officers preferred pistols to revolvers until they found out that the Prussians were using revolvers.) At this point, according to *Oruzheinyi Sbornik*, Gorlov presented for consideration an improved Smith and Wesson revolver. News that the revolvers had received high marks in American tests impressed the small arms committee, and on 1 May 1871, the first contract for twenty thousand revolvers was signed in Springfield.[58]

The sale of Smith and Wesson revolvers to Russia generated a mutually beneficial relationship between the Springfield company and the Russian government. While Russia received a superior weapon made with interchangeable parts, Smith and Wesson benefited technically and commercially. In the 1850s, the Russians had shown preference for the self-cocking revolver developed by Adams and Deane in England over Colt's single-action revolver. According to company historian Roy Jinks, Smith and Wesson designed a self-cocking revolver, "either at the request of the Russian government or in anticipation that there would be a Russian market for a self-cocking style handgun." While the U.S. Army was slow to adopt the company's newest designs, Jinks argues, the Russian government was the first to adopt the model 3 metallic cartridge revolver, considered the most advanced in its time. Moreover, Gorlov and his successors, K. Ordinets and N. Kushekevich, worked in Springfield as designers, supervisors, and inspectors of the Russian model. To a former president of the company it seemed that "the shop soon was swarming with overzealous Russians and finally an exasperated Daniel Wesson called a halt and insisted that the 'technicians' be returned to the land of the czars." Yet there is little evidence that the Russians hindered operations in Springfield. Jinks, in a more sober assessment, regarded the Russian officers, who insisted on only the best, as highly demanding and meticulous. The Russians were much more demanding than Turkish or even American inspectors: Smith and Wesson routinely rubbed Russian markings off guns rejected by the Russian officers and then sold them on the American market. The company put up with the demanding Russians, Jinks reasons, because it respected them and their expertise.[59]

The Russian contract brought more than commercial or technical advantages for the Springfield Company. Just as Colt's foreign sales generated publicity for his revolver, Smith and Wesson's foreign sales brought prestige. While Colt had had several foreign orders and a factory in England before his musket sales to Russia, for Smith and Wesson, the Russian contract was their first major foreign order. To take one example, the fact that the revolver "has been extensively adopted for the Russian government by its calvary" was "evidence of superiority" in the eyes of Norton.[60] Gorlov himself, of course, would have concurred. In the context of a discussion concerning payment for parts, Gorlov wrote to Wesson:

> I hope that you will be kind enough to instruct Mr. King not to make too large a margin in this last estimate, so as to allow us to work in the future hand in hand to improve your pistol and bring out such a model that will ensure for it the power to drive away any competition here as well as in Europe.[61]

According to Jinks, the Russian contract "meant the world" to Smith and Wesson. It improved American sales, got Smith and Wesson out of the financial difficulties of the late 1860s, and established the company as a major producer in Europe and around the world.[62]

Gorlov calculated that the company made 142,333 Russian contract guns between 1871 and 1881, and another 75,000 made by license were sold to the firm of Loewe and Loewe in Berlin. Smith and Wesson remained the Russian service revolver until the adoption of the Liège-made Nagant in 1895.[63] Opinion of it, particularly regarding its superior ballistics qualities, design, and durability, remained high throughout the 1870s and the 1880s. By the end of the 1880s, however, more and more criticism of the revolver, particularly due to its heaviness, was expressed. *Oruzheinyi Sbornik* summed up the later attitude toward the Smith and Wesson revolver:

> Generally speaking, the Smith and Wesson revolver, as many years of use have shown, is a superior military arm. But even the sun has spots and the same is true of the Smith and Wesson.[64]

Gorlov had one more mission to fulfill while in the United States. In 1867, Richard Gatling, the inventor of a prototype of the machine gun, sent two models of his gun to Russia. While in Hartford, Gorlov was instructed to modify the gun to Russian specifications, in particular to modify the chamber to use the .42 caliber metallic cartridge already planned for the Berdan rifle, which would be "a most useful provision in case the supply of either one or the other should run short." In addition, Gorlov increased the number of revolving barrels from six to ten and doubled the rate of fire to two hundred rounds per minute. In 1871 the Russian government purchased four hundred of the weapons manufactured by Colt as well as a license to manufacture the improved "Gorlov gun" at the Nobel plant in St. Petersburg.[65] A new and unpredictable weapon, the Gatling gun was used sparingly, although it did prove successful in the various Central Asian campaigns. If nothing else, the odd-looking weapon, which could be carried by camel, was a conversation piece: "The Russian mitrailleuses were ordered in a panic after the outbreak of the Franco-Prussian War, and so great was the enthusiasm when they arrived at St. Petersburg, that processions of welcome were organized by ecstatic ladies."[66]

Russian Rearmament: An Assessment

In 1872 the Ministry of War considered that the committee for rearmament had successfully completed its tasks and transferred its duties to the Artillery Department. Colonel Gorlov was promoted to the rank of major general. Thus ended a short period of intense effort to modernize Russia's infantry arms. In an attempt to find the right breechloading mechanism, from 1866 to 1870, the army adopted rifles at an average of one new system per year. Given that the major European armies themselves had only recently adopted breechloaders, Russia could no longer be considered backward in firearms design. In 1871 two visiting American artillery officers were "favorably impressed with the great progress made in Russian artillery."[67] Equally impressed was the architect of Russia's military modernization, Dmitrii Miliutin. Although adoption of new systems and actual supply to the regiments were two very different matters, the minister of war was confident that the infantry could now be armed with modern weapons during wartime. He later looked back on the rearmament of the late 1860s with a certain amount of satisfaction:

> Since 1866 we have changed the entire weaponry of our infantry . . . first from smoothbore muskets . . . to rifles . . . then to rapid-fire arms. . . . In 1860 we had but 216,000 muskets . . . in our arsenals. At the present time we have more than one million rapid-fire rifles.[68]

The particular needs of the Russian infantry and the productive capacities of Russia determined the choice of the new arms technologies and the pattern of adoption of new weapons systems. Because of the large, and largely peasant, infantry and the geographical dispersion of the Russian army, the official infantry arm had to possess two essential military features—durability and convenience of maintenance. Repeatedly in the military sources, these features were placed ahead of ballistic features such as rapidity of fire, range, and accuracy in appraisals of the advantages or disadvantages of any given system. The shortcomings of the nation's productive capacities dictated a third essential feature: simplicity of manufacture. In his 1862 report, Miliutin recognized this two-pronged problem:

> Foreign artillery technology has diversified and improved so rapidly in recent years, that given the size of our army, we have found it difficult to settle on one system, and given the limited capacity and low development of our engineering works, we have been obliged to proceed with extreme caution in adopting innovations. This is one of the reasons why, despite the most energetic measures of the artillery department and despite the diligence and expertise of our artillery officers, we must at last admit that in material and arms of our army, we are behind other European countries.[69]

In the early stages of rearmament, Russia borrowed weapons innovation—rifling the bore, changing the lock mechanism, loading at the breech, for example—as well as system design from abroad. Thus, the 1857 infantry rifle, the Terry-Norman percussion breechloader, the Carl needle breechloader, and the Krnka metallic cartridge breechloader were all essentially of foreign design. At this point, rather than purchase the weapons abroad, Russia made a dogged attempt to produce its own conversion models. But Russia's overall backwardness, particularly, as Miliutin suggested, in engineering and machine-building, made the process of conversion in the 1860s difficult and explains the thoroughness and caution of the small arms experts in choosing new systems. To make matters worse, this stage of rearmament coincided with a period of administrative instability at the government armories. Consequently, establishing the workshops necessary for conversion proved to be costly and time consuming. No sooner were the armories prepared to manufacture one conversion model than an improved model appeared. Conversions were merely a temporary solution; new systems offering rapid fire had to be developed and supplied. As the editor of *Oruzheinyi Sbornik* argued, somewhat ironically, because of a shortage of skilled mechanics in Russia, the only way to speed up rifle conversion was to adopt a simple system using metallic cartridges. It seemed most rational to take advantage of experience already gained in foreign countries; that is, to take advantage of backwardness and send military agents abroad. The decision about where to send agents was determined by design and production considerations. As one artillery officer wrote in the midst of the feverish activity in rearmament,

> Moreover, one must consider that only America, with its myriads of new arms and cartridges factories, which operate with the latest machines, could facilitate the constant conversions and improvements in the search for a model and cartridge that had to be faced at every step by trial and error.[70]

Once the decision had been made in 1866 to adopt the metallic cartridge, system selection was determined by a somewhat unusual process: new rifles, carbines, and revolvers were designed to fit the latest developments in cartridges. The Russians had developed a system of interchangeability of cartridges. Once this decision had been made, it was natural to turn to the United States, generally recognized in the post–Civil War period to be the leader in the development of metallic cartridges. This interchangeability of cartridges greatly facilitated subsequent design and production decisions, and in the later stages of rearmament, therefore, the pattern of weapons adoption changed somewhat. Innovations in system design proceeded far away from Russia but, paradoxically, Russians played a greater role than had been the case earlier. Once a new system had been tested and adopted, the government placed initial orders in the United States and, to a lesser extent, in England. Thus, the Berdan-Gorlov metallic cartridge, the Berdan-Gorlov rifle, and the Smith and Wesson "Russian revolver" were designed and initially produced in America. The Russian government ordered relatively few guns abroad. The interchangeability of cartridges considerably simplified design and, even more important, promised to simplify manufacturing.

The Russian government had decided that it was not content merely to purchase arms from the United States, even if they were manufactured at the best factories. Russia was determined to make its own arms. Just as system design could dictate subsequent production decisions, the reverse was also true. Comparing the two Berdan systems, N. Litvinov admitted that considerations concerning the ability to manufacture a system in Russian armories carried greater weight than considerations concerning the purely technical qualities in system selection. The superiority of the Berdan 2 over the Berdan 1 "was not so much in its military capabilities as its promised superiority for large-scale production at our armories."[71] Litvinov stated that, after system selection, limited orders could be filled quickly and satisfactorily from abroad. But the long-range goal was simplicity and ease of domestic manufacture. Ironically, Gorlov, a proponent of the Russian Berdan 1 became a critic of the Berdan 2; instead he expressed a preference for the military qualities of the British-made Henry-Martini. However, concerning the adoption of the Berdan 2, both Gorlov, angry with his superiors for not taking his advice, and Soviet historians of the 1950s, who accused the Russian government of simple-minded attraction to everything foreign, seem to have missed the point. Replying to a memorandum from Gorlov, Minister of War Miliutin stated:

> If Mr. Gorlov was referring to the convenience of ordering the Henry-Martini rifle in England, this factor alone can hardly justify another change in model. *Russia is not Egypt, she is not the Papal States, to limit herself to foreign orders to equip her army.* We must build our own factories so we can make our own weapons in the future. But this raises a crucial question: Will our factories not encounter insurmountable difficulties in making the 1868 model rifle [Berdan 1] even if we agree with Gorlov that these rifles are the best available? The Berdan 2 model is appealing especially because of its simplicity of fabrication and the ease in handling its bolt.[72]

It is clear that the imperatives of domestic production weighed heavily in rearmament decisions. It might seem from Miliutin's statement that the value of the new technology of rapid firepower was universally accepted and that the only obstacles to adoption and deployment of new systems were economic. In fact, Russian officers debated not only the technological and economic feasibility of new breechloading rifles and metallic cartridges but also the effect of these changes on military strategy and tactics and on the training and morale of the soldier. This debate, clearly evident in the pages of military journals, is significant because it took place in the context of the larger discussion of military reforms—during the last of the major Great Reforms—and also on the eve of the Russo-Turkish War.

Firearms in Russian Tactical Doctrine After Crimea

The increases in firepower brought about by the new weaponry of the nineteenth century posed a challenge to military tacticians and commanders. Loose, extended, skirmish-line tactics became an alternative to the assault of infantry columns in closed formation; the cavalry charge came to be viewed as a suicidal tactic; and open formations placed a premium on individual initiative. While adherents to the old ways survived and while mid-century wars had their share of suicidal cavalry charges, the nineteenth century had become the age of the rifle. Given that Russia was undergoing rapid changes in its social, legal, educational, and military structures, and given that rearming the infantry with modern weapons became a high priority in the 1860s, a corresponding reexamination of tactical doctrine and training is not surprising.

One might expect such a reexamination to come from the top down, as so much else did in Russia in those days. According to Bruce Menning, Dmitrii Miliutin was a proponent of tactical flexibility and of rapidity of calculation and action. To accept responsibility and to seize initiative were qualities Miliutin admired in an officer. And, as a vigorous proponent of a literate population, Miliutin admired these qualities not only in officers but also in men.[73] Although these principles have a certain modern ring, they describe the sterling qualities of a commander in any age and do not necessarily translate into a particular tactical doctrine, let alone a tactical doctrine that takes into consideration firepower.

One tactical doctrine that recognized the importance of firepower was that of General A. I. Astaf'ev. In *On Contemporary Military Art*, written immediately after the Crimean War, Astaf'ev began with the tactical assumption that the infantry column could no longer attack against fire. Several principles followed this assumption. Columns must be mobile in order to move away from enemy fire. Mobile columns, in turn, suggested a merging of the functions of the infantry of the line and of light infantry. Maneuver in dispersed formation accorded freedom of movement and speed and facilitated the quick flanking operations necessary to protect against fire. The infantry attack commenced in dispersed formation and then proceeded to the closed formation for the shock of the mass. General Zeddeler also advocated open order and the independent skirmish line. In sum, Astaf'ev and Zeddeler went against the preponderance of Russian tactical doctrine by declaring that fire was more important than bayonet.[74]

However, as an account of the Russo-Turkish War later stated, "The prevailing ideas in military circles and in the professional literature were not conducive to utilizing the distinguishing qualities of contemporary firearms, distances, and rapidity of fire."[75] The bayonet school continued to dominate Russian tactical doctrine, and the most consistent proponent of the bayonet school during the reform era was M. M. Dragomirov, professor of tactics and later head of the Nicholas General Staff Academy. Dragomirov wielded considerable influence among officers during the reform era, was a major spokesman of the "back to Suvorov" movement, and his 1879 textbook on tactics remained the standard text for officers until the Russo-Japanese War.

The root of Dragomirov's preference for the bayonet was philosophical, not tactical or technical. To Dragomirov, the firearm and the bayonet corresponded to two different sides of human nature. The firearm, with its premium on accuracy and skill, represented reason, while the bayonet represented bravery and will. Dragomirov believed that man's essence lay in the latter and that in battle the most important factor was human. It followed then that the bayonet held supremacy over the firearm. Dragomirov believed that the firearm, no matter how perfect, was an extension of the lethal capacities of the soldier (*"Deistvuet ne oruzhie, a chelovek"*). To use the bayonet, the soldier "does not need much thought or imagination, actually, the less thinking the better."[76]

Dragomirov, who was not opposed to employing breechloading rifles, drew a distinction between the approach to combat and the final attack. "The bullet and the bayonet do not exclude but complement each other; the former paves the way for the latter."[77] Cold steel at close range would then give a quick and decisive victory. Yet Dragomirov feared that improvements in firearms would lessen the significance of cold steel. To begin with, Dragomirov argued, since the soldier must have the conviction that he is invincible, reliance on firepower might lull the soldier into a dependence on his weapon rather than on his own valor. Concern for finding cover in order to fire made the soldier preoccupied with his own personal safety. "One does not need much intelligence to expose oneself to the bayonet; likewise one does not need much courage to fire at the enemy from behind a rock or a bush at a distance of 400 to 500 yards, let alone of 1,000 yards"[78] Second, as Bruce Menning points out, preoccupation with rifle fire might cause the soldier to become overly concerned with marksmanship, which requires thinking. But too much thinking leads to excessive individuality and loosens and demoralizes the mass. Third, Dragomirov feared that excessive reliance on firepower would undermine discipline. In closed formation, "the officers and certainly the soldiers, are machines, giving blind, almost unconscious obedience to command." In open formation officers and men alike had to choose for themselves the target, the best position, and the appropriate time to open fire. Rifles thus lessened the officer's control over his men. This placed too great a premium on thought and independence. "Little by little the men are offered more independence than that which should be offered a soldier."[79]

Dragomirov favored the tactics of maneuvers in closed formation over those in open formation. Closed formation, characterized by a unified "shoulder to shoulder" movement, meant a primacy of the mass and a denial of the individual and of the right of the individual to think. Open formation, characterized by skirmishers firing at greater distances and at varied times, denied the mass and emphasized the individual's right to think.

> Improvements in the accuracy of firearms, by raising the issue of respect of the individual, dismiss the significance of form, and give too much room for man's egotistical impulses. The desire to have accurate firearms must be curbed even more with improved firearms, and it is absolutely essential that marksmanship be balanced with maneuvering.[80]

Dragomirov labeled excessive reliance on firepower "fire-worshipping" (*ognepoklonnichestvo*). Although it might be appropriate for European armies to emphasize the soldier's intelligence, initiative, marksmanship, adaptation, and action in dispersed formation, Dragomirov argued, the Russian soldier must display courage, endurance, strength, and duty. These qualities—in short, the will to fight—were more decisive than the training or the use of the weapons.[81]

General G. A. Leer (1829–1904), professor of tactics, chair of strategy, and Dragomirov's successor as director of the Nicholas General Staff Academy, echoed this attitude. Victory depended on moral force, firepower was preparatory, and the bayonet assault in deep order was decisive. Such thinking influenced, for example, Leer's evaluation of the Prussian needle gun and of Prussian military successes in the 1860s: the alleged merits of the Prussian needle gun were of little account; success was a result of superior generalship and of a preponderance of well-trained forces.[82] This emphasis on bayonet over bullet in Russian tactical doctrine, according to American historians Peter Von Wahlde and Bruce Menning, showed an overestimation of the efficacy of cold steel and an underestimation of the importance of firepower, and delayed realization of the impact of modern firepower on tactics.[83]

The prevailing tactical doctrine was reflected in the regimental organization of the 1860s. In an excellent study of Russian tactics during this period, Menning shows that assault formations, by their very organization, emphasized cold steel. Each infantry battalion consisted of five companies—four companies of the line and one company of sharpshooters. In the battalion attack formation, the four companies of the line advanced in closed order to deliver the bayonet assault behind the skirmish line established by the fifth company of sharpshooters. "This prescribed assault formation," claims Menning, "placed four-fifths emphasis on cold steel and one-fifth on firepower." In actual practice the emphasis on fire was reduced even more. Although the new infantry regulations of the early 1870s recognized the importance of the dispersed formation and of the soldier's initiative to choose targets, in practice, according to Menning, "commanders attempted to keep a tight rein on the skirmish line. . . . An obsession with orderly formation caused commanders to resist reinforcing the skirmish line."[84] The Russian soldier was not accustomed to acting on his own initiative, and the long skirmish line did not permit officers to direct men by voice and example. In the words of one observer, "the prejudice still exists in high and influential quarters in favor of the old bayonet tactics with steady shoulder-to-shoulder movements, in opposition to new-fangled theories about fighting in extended order and the development of rifle fire."[85]

The tactical doctrine of the bayonet school was also reflected in training. To begin with, universal conscription in 1874 did not bring with it universal education, which was allegedly inappropriate for the Russian peasant. Although poor performance in the Crimean War was occasionally blamed on the lack of training and practice in marksmanship, requests for extra marksmanship training were repeatedly denied and volley fire by command continued. Moreover, prejudices against alleged ammunition waste also prevented greater attention to marksmanship training. This was one of the most frequent arguments against rapid-fire weapons: "The rifleman, in the confusion of battle, may fire all his ammunition at once and become defenseless."[86] The soldier might expend his ammunition firing at long range, leaving him with insufficient ammunition to fire at close range. It was argued that ready-made cartridges, which could be wastefully expended, were not appropriate for officers in battle. Suvorov's dictum "Wait—and fire straight," referring not only to attack but also to defense, was the prevailing wisdom regarding fire.[87] The military historian Fedorov compared the training of the Russian soldier unfavorably with that of his European counterpart:

> In the West, the soldier is trained to act on his own, while in Russia soldiers act in mass, do not take initiative, and are inflexible. When rifled arms were introduced in European armies, much attention was given to the soldier's flexibility, mobility, agility, action in open formation, and adaption to surroundings. At the same time the Russian army, raised on the scriptures of Suvorov which put bayonet force over accuracy of fire from long range, ignored the individual training of the soldier and the rapid adoption of rifled arms.[88]

It would seem that the average Russian private was not to be trusted with sophisticated weapons. According to Pototskii, the Russian army did not want soldiers to have a new weapon unless they had had at least one year's experience with it. Before the Turkish War only specially trained units were armed with, or received training in, the Berdan rifle. It was not until 1879 that the Berdan was introduced generally and the entire infantry was placed on the same footing as the sharpshooters. As a result, the infantry of the line was not properly trained in the use of the Berdan:

> Although marksmanship has been taught rather assiduously, and ever since 1871, it has been checked by inspections, it has been taught chiefly at distances no greater than 600 yards; mass fire for tactical purposes has never been used. Not only were long-range and rapid fire rejected, but from closed formation, only volleys, not individual fire, were permitted. Regarding the general tactical training of the infantry, one could say that despite the technical improvement in small arms, we have viewed firepower as supplementary and have considered the strike of cold steel the most important. Volleys from closed formation are emphasized, while single target fire has been considered merely a supplement.[89]

Although traditionalist Russian officers can be faulted for their adherence to old ways and prejudices, there were some very practical bases to their skepticism regarding firepower. To begin with, extra marksmanship training was impeded by shortages of good rifle ranges and instructors. More important, the fear of excessive use of ammunition, whether in training or in the field, was grounded in the fiscal stinginess of the Russian army. Profligacy in ammunition could easily result in unacceptably high expenditures. Finally, it must not be forgotten that rearmament in the 1860s and the 1870s took place in the context of serious reservations about the new arms. Reservations were based in part on a skepticism of the technical efficacy of the new arms in battle and of the inappropriateness of breechloaders in the hands of untrained troops. This skepticism, shared by many European officers, was based on concrete difficulties in using what were, after all, new weapons.

This skepticism was particularly widespread concerning the revolver. Russian officers argued that the new weapon had important technical drawbacks. At the moment that Colt revolvers were first being examined, Major-General K. Konstantinov concluded, "While granting full recognition to the cleverness and in certain cases to the utility of the revolver, it must be said that this arm, due to its lack of durability and the difficulty in maintaining it, lacks the most necessary qualities of a military arm."[90] Later writers such as N. Egershtrom believed that revolvers could not replace pistols as a military arm because of their complexity, high cost, and difficulty of assembly and maintenance during wartime. A. Vel'iaminov-Zernov, author of a textbook on firearms, claimed that revolvers were highly unreliable because "quite frequently at a crucial moment the revolver betrays its calling and makes firing dangerous. For example, malfunction of the mechanism (which indeed occurs often and is impossible to anticipate) may cause the chamber to line up imperfectly with the barrel."[91] According to M. Terent'ev, reloading the revolver left the cavalryman defenseless, and accuracy depended mainly on the stability of the horse: "Indeed the excessive distribution of firearms to the cavalry causes more harm than good, because it in fact facilitates the replacement of cold steel, the essence, the specialty of the cavalry, with the firearm; in short questionable success is preferred to certainty." Accordingly, the revolver had only limited usage for detachments executing a specific mission—storming a fortress, seizing a ship, quelling a rebellion.[92]

Despite the skepticism, the new weapons had vigorous defenders. One of these defenders was Alexander Gorlov. Gorlov, of course, was influenced by his experience in America and by the attention given rifles during the Civil War. "The war in America," Gorlov wrote in 1866, "demonstrates convincingly that it would indeed be dangerous to enter a war today without rapid-fire arms." Breechloaders were used effectively in repulsing or launching an attack and by soldiers not accustomed to handling firearms at all. Responding to a common accusation against rapid-fire arms—the possibility to waste ammunition—Gorlov observed that "trusting his firearm, the soldier is much more at ease; he does not rush to fire as used to happen with old-model arms." Moreover, Gorlov argued that because a battle involving breechloading rifles was decided more quickly and because there were long periods when no shots were fired at all, cartridge use was not any greater. Finally, by forcing the soldier to "give more attention to the care and maintenance of expensive weapons," the new breechloaders could serve an instructional purpose.[93]

Philosophical and tactical antipathies toward rapid-fire small arms, the lack of training in their use, the alleged propensity of the soldier to waste ammunition, and the improper care, maintenance, and repair of complex weapons all worked against the rapid distribution of breechloaders and revolvers in the army. Although the Ministry of War moved quickly in the late 1860s in its quest for the best conversion system and in its adoption of breechloading rifles and metallic cartridges, actual procurement, distribution, and deployment of new systems proceeded slowly.

Reports of delays in distribution of new weapons abound. Although the Ministry of War claimed in 1865 that all infantry divisions were armed with .60 caliber rifles, according to the Artillery Administration, many Cossack divisions were still armed with smoothbore muzzleloading muskets. According to British Foreign Office records, at the beginning of 1872 the Russian Ministry of War claimed to have enough breechloading rifles for the army to be on a war footing; that is to say, 860,000 breechloaders.[94]

Although the total number of arms may have been impressive, proportions told a different story. Of 860,000 breechloaders, only 30,000 were the modern Berdan 2 rifles adopted in 1870. Equipping the army with Berdan 1 and Berdan 2 rifles was gradual and began with the sharpshooter battalions. No muzzleloaders were converted on the Carl and Krnka systems after 1873; yet through the 1870s battalions of the line, local forces, and militia units still had the .60 caliber Carl and Krnka rifles. By 1 January 1877, the guard, grenadier, and all sharpshooter battalions had the Berdan 2; other infantry divisions had the Carl and Krnka. Besides the sharpshooter battalions, only 16 of 48 infantry divisions were armed with the Berdan 2, despite the fact that the infantry was supposed to have received the Berdan 2 at the outbreak of the Russo-Turkish War. Although an estimated 230,000 Berdans were in reserve supply, they were not distributed to the infantry. Indeed, while the entire army was scheduled to be rearmed by September 1878, it was only in 1884 that rearmament was completed, by which time the Berdan was already out of date.[95]

The slowness in the supply of new weapons might have been less noteworthy had it not been for the outbreak of war with Turkey in 1877. Although the Minister of War may have judged the army prepared for war in its small arms, it was clear that it was not. The Turkish army was arguably better-armed with new weapons. The Turkish government in 1873 awarded a contract for 600,000 rifles to the Providence Tool Company. These rifles were breechloaders of the Peabody-Martini system using a rim-fire metallic cartridge, a version of which had also been ordered by the Swiss, French, and Rumanian governments. Although it had been tested frequently in the United States and Europe (including Russia), no European army adopted it until the Turkish army did so in 1873. With a range of 1,800 yards and the capacity to fire 30 cartridges per minute (compared to the Berdan's 20 cartridges per minute), the Peabody-Martini was a potent weapon. In addition to using rapid-fire long-range rifles, the Turks ordered 39,000 Henry-Winchester repeating rifles and were the first to use this magazine system in battle.[96]

It should not be surprising, then, that Russian soldiers were struck down at two thousand yards, according to Captain Kuropatkin. Divisions numbering between ten thousand and twelve thousand men were quickly reduced to four thousand men. It should be noted that the alleged superiority of Turkish arms was due to the fact that frequently in battle Turkish Peabodys faced Russian Krnkas. A noncommissioned officer of the Vladimir regiment, two thousand yards from the Turkish redoubt at Plevna, described the loss of officers and the demoralization that resulted from long-range fire:

We had not been long in the vineyard when the Turks began to fire at us. Many of our men were wounded before the order was given to advance . . . and among them the captain of my company who was lying down among the vines. . . . Men fell on all sides, in the front ranks and in the rear sections alike. . . . When at last we moved forward, the bullets fell upon us like hail. . . . We had not gone more than 50 paces when the officer of my subdivision was struck in the chest. We could not fire. Our Krnka rifles were only sighted up to 600 yards and the Turks were a verst and one half away. Long before we got near the [Turkish] trenches there was no one left to advance.[97]

Russian officers were well aware of the capabilities of the Peabody, and some Russian small arms experts rated it over the Berdan. Nevertheless, the memoirs of the Russo-Turkish War are rich in praise of the Berdan rifle. The most extensive description comes from V. F. Argamakov. Writing his memoirs in 1911, Argamakov recalled that the Krnka's poor extractor mechanism meant that most soldiers were "defenseless after the very first volley" while they tried to expel the cartridge case with a ramrod. "The Krnka more often than not turns into a club, impeding the soldier like an extra bayonet. I abandoned hope when I heard the soldiers' curses and saw with my own eyes that they were justified." Later, Argamakov, who referred to the Berdan as "outstanding," an arm "superior in accuracy and in range," a "pearl," and even "heavenly," continued, "Rumors spread that we were going to get Berdans. Well, praise the Lord! If only the soldier could have a Berdan and some cartridges, the Berdan won't leave him defenseless."[98]

On the eve of the Russo-Turkish War, a major in-house study of the transformation in the Russian military concluded, "Having adopted [rapid-fire] arms, it was necessary not only to change weapons but also to accede to fundamental changes in training and tactics; the old ways, having had an honored tradition, stood in the way of new ways that were not yet proven in battle."[99] Russia's inability to coordinate fire and movement, its incorrect positioning of guns, its adherence to closed formation, its ignorance of the impregnability of positions defended by breechloaders, and the lack of individual initiative on the battlefield spelled disaster at Plevna. In Menning's estimation, the Russo-Turkish War exemplified excessive reliance on speed and morale and the neglect of firepower.[100] Argamakov summed up the criticism of Russian tactics:

> Therefore, not having thought of ways to use this superior weapon and not having fully utilized the sharpshooters for what they were intended, we treated our Berdans as mere cudgels, with which our sharpshooters and the guards led the attack, hoping that they could hold onto this cudgel until they engaged in hand-to-hand combat. . . . We forgot that the utility of something should not be rejected without understanding the means of using it. . . . With all the foresight of a chicken, we did not recognize our Berdans for being the pearls they were.[101]

It has been argued that, "faced by the increasing complexity of modern weapons and tactics, conservative commanders retreated morally and mentally to the shock weapon they could understand, that seemed both safe and heroic, and intimately connected with the social myth by which military castes in Europe maintained political power and economic security."[102] Thus the cult of "cold steel" that preserved the bayonet as the premier shock weapon was closely bound with the cult of courage, discipline, and will power. Nowhere was this more pronounced than in Russia. Relying on the bayonet, officers for a long time were unwilling to recognize the profound changes brought about by firepower. They feared the loss of control over soldiers trained to take the initiative and to think in battle. They feared that soldiers armed with breechloaders would exacerbate chronic supply problems by wastefully expending ammunition in battle. However, the persistence of the bayonet school in Russian tactics provides only a partial explanation for the small number of Russian small-caliber rifles on the battlefield in 1877 and 1878 and for their slow integration into tactical doctrine. Russia, unlike its opponent, was not filling its needs by massive orders from foreign arms makers. The bulk of its needs had to be met by domestically produced weapons, the production of which had only recently been organized. The temporary difficulties associated with wartime production orders compounded already trying times for the government armories.

LABOR, MANAGEMENT, AND TECHNOLOGY TRANSFER

The Reorganization of the Russian Armories

"Russia is not Egypt, she is not the Papal States, to limit herself to foreign orders to equip her army. We must build our own factories so we can make our own weapons in the future."
—*Minister of War Dmitrii Miliutin*

● GIVEN THE LONGTIME COMMITMENT OF Russian officers to the bayonet school, it is understandable that firepower was subordinated in tactical doctrine, in training, and on the battlefield. However, discussion of the new rifled breechloaders and revolvers had another dimension. The infantry could be supplied with small arms mass-produced with standardized parts in giant mechanized armories, thereby speeding the service and repair of arms on the battlefield. Import of arms from either traditional suppliers in Liège or new suppliers in America still left Russia behind in the advanced technologies of the day. Moreover, Russia's native machine-tool and machine-building industries were incapable of responding quickly to the increased demand for the new technologies of mass production of standardized weapons.

Accordingly, in addition to small initial orders of new weapons, the Russian government imported the new production technologies and began outfitting small arms and amunition factories with machinery. Over the course of the next decades, the Russian government fostered the manufacture of the new weapons in Russian small arms factories. This process attempted to domesticate foreign technology and reorganize the work processes in the small arms industry.

Labor Force

Since the time of Peter the Great, the corporation of gunmakers was indentured to the three government armories and obligated to fill state weapons orders. The armory administration, for its part, was responsible for delivering quotas to the state and for providing employment to the gunmakers. The resulting administrative, economic, and social system fit well with the state's needs for military orders and for the social order. That this system did not serve the needs of labor productivity, technological innovation, or industrial progress—an argument not infrequently made during the first half of the nineteenth century—was of secondary import given the prevailing social system of serfdom and the slow pace of technological change. However, once that social system no longer had a raison d'être, the system of indentured labor at the government armories was challenged.

An Artillery Department Commission created in 1856 to study the government armories concluded that the Crimean War demonstrated the backwardness of the Russian arms industry and criticized several aspects of the system of serf labor as "constituting a perpetual obstacle to our improvements in firearms." To begin with, the system of indenture and labor recruitment severely limited the sources of skilled labor in the gun trade. Second, neither labor nor management had the incentive to improve production and to adjust to mechanized procedures. The labor force possessed job security but lacked the training and inclination to respond to innovations. Management, obliged to provide for the livelihood of workers, had an important disincentive to mechanize production. According to the commission, "The introduction of machines, which have quickened and improved labor, became infeasible: they have put many workers out of work and, consequently, out of a livelihood. The machines that have been acquired, viewed as a hindrance rather than a benefit, operate only to keep the worker employed and have been carelessly maintained."[1] Finally, the serf gunmakers were allegedly unproductive, untrustworthy, and dishonest: "It would seem to be generally indisputable that only free men are capable of honest work. He who from childhood has been forced to work [*rabotal iz-pod palki*] is incapable of assuming responsibility as long as his social condition remains unchanged."[2] Therefore, concluded the commission, the mechanization of the government armories required the abolition of indentured labor. Six years later, in his famous report outlining proposed changes in Russia's military, the Minister of War, Dmitrii Miliutin, repeated the need for the abolition of indentured labor.[3]

On 19 February 1861, the tsar signed the edict emancipating the proprietary serfs. This was followed by the emancipation of other categories of the serf population, including state peasants in 1866. Most important in this discussion is the emancipation of the Tula gunmakers in 1864 and of the Izhevsk and Sestroretsk gunmakers in 1866. At all three armories, the gunmakers were freed from obligatory labor. The Tula gunmakers were placed in the *meshchane estate*, while their brethren at Izhevsk and Sestroretsk were granted the status of villagers and given the same rights as the newly emancipated state peasants. At all three armories, the state lands on which gunmakers had been living became private property: gunmakers retained their homes and garden plots as private immutable property; arable land became the communal property of the gunmakers' association as a whole; and households received grazing land as an allotment. Tula gunmakers who had been registered in the association for at least twenty years were exempted from state taxes and from military conscription for life; those who had been registered for less than twenty years were granted these exemptions for a period of six years. At Izhevsk and Sestroretsk, gunmakers who had worked at the armory for at least twenty years could receive a passport from the township office and leave.[4]

A few observations can be made about the emancipation settlement and the labor force in the Russian arms industry. First, as in the case of emancipated serfs, the government regarded ties to the land and to a communal association to be the bases of the social order. Although it may have made sense for the government to encourage emancipated serfs to retain land and remain farmers, it would seem less economically rational for armorers to continue to be landowners and part-time farmers. Second, the government adopted a somewhat inconsistent policy regarding gunmakers with seniority. By bestowing the greatest privileges on senior Tula gunmakers, the government clearly expressed a desire to reward and retain its most skilled workers. At Sestroretsk and Izhevsk, in contrast, senior gunmakers could receive a passport and leave. Many did just that. At Izhevsk, for example, at the beginning of 1867, 4,125 gunmakers worked at the factory; by April, 2,637

Colt 1851 Navy model revolver made in Izhevsk. In 1854 Colt sold a license to make Colt revolvers at the Tula and Izhevsk small arms factories. The Russian markings on the barrel show the date, 1856; a double-headed eagle, the symbol of Imperial Russia; and the name of the factory, "Izhevskii zavod." Collection of Al Weatherhead; photographs courtesy of Al Weatherhead.

Model 1851 Navy revolver, in wooden case, presented by Samuel Colt to Grand Duke Constantine in 1858. Collection of the Hermitage Museum in Leningrad; photograph courtesy of Leonid Tarassuk and R. L. Wilson.

The Smith and Wesson Model 3 Russian First Model revolver. The Russian government contracted with Smith and Wesson for 20,000 revolvers when the revolver was adopted as the official army revolver in 1871.

The Smith and Wesson Model 3 Russian Second Model infantry revolver. The Russian government ordered 70,000 of these revolvers from Smith and Wesson.

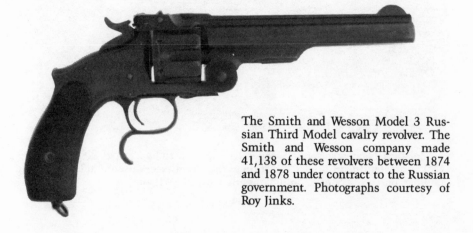

The Smith and Wesson Model 3 Russian Third Model cavalry revolver. The Smith and Wesson company made 41,138 of these revolvers between 1874 and 1878 under contract to the Russian government. Photographs courtesy of Roy Jinks.

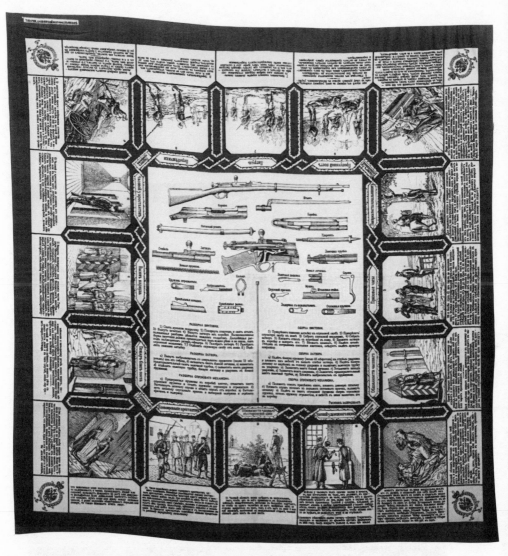

A ceremonial army kerchief made by the Danilov mill in Moscow. The center shows the parts of the Berdan 2 single-shot bolt-action infantry rifle along with instructions for assembly and disassembly. Around the edges are 16 vignettes of military life. The cloth was made for the All-Russian Exhibition of Science and Industry held in Moscow in 1872. Courtesy of the Library of Congress.

Berdan 1 single-shot trap-door action infantry rifle. The rifle was adopted by the Russian army in 1868, and 30,000 rifles were made for the Russian government by the Colt Company in Hartford. The close-up view shows the name of the Colt Armory, Hartford, in Russian. Photograph courtesy of the Division of Armed Forces History, National Museum of American History, Smithsonian Institution.

Berdan 2 single-shot sliding bolt action infantry rifle. The rifle was adopted by the Russian government in 1870 and was sometimes referred to as the 1870 model infantry rifle or, colloquially, "the Berdanka." Although initial orders for 30,000 Berdan 2 rifles were placed with the Birmingham small arms company, more than 1,000,000 were eventually made in the Tula, Izhevsk, and Sestroretsk armories. The close-up view shows the barrel markings indicating that the rifle was made in 1881 at the Sestroretsk Armory. Photographs courtesy of the Division of Armed Forces History, National Museum of American History, Smithsonian Institution.

Detail of the Berdan 2 infantry rifle showing the bolt-action breech-block in closed and open positions. Drawing back the bolt handle to its full extent opens the breechblock and cocks the rifle. Opening the breechblock compresses a spiral mainspring that when released through the trigger is impelled against the firing pin, discharging the ammunition. The bolt action allowed a greater rapidity of fire than was possible with the trap-door action of the Berdan 1 model.

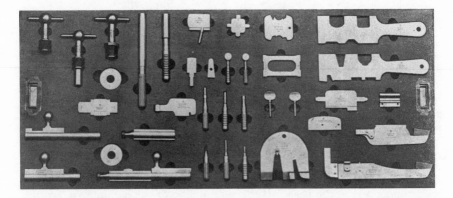

A set of gauges used to inspect the bolt of the Berdan 2 infantry rifle. These gauges were part of the machines and equipment sold by the Leeds engineering firm of Greenwood and Batley in 1871 to the Russian government. The gauges, which allowed an inspector to tell whether a specific part of the rifle was finished correctly, were essential for making weapons with interchangeable parts and for quality control. Photograph courtesy of William B. Edwards.

A die-forging machine at the Tula Armory in the early nineteenth century. Before 1819, lock parts were forged manually by the gunmakers at home with simple blacksmith's tools, a process that was inaccurate and necessitated complicated filing and fitting. Later, John Jones in England devised a machine for forging, or stamping, the lock parts by dies. The greater accuracy of parts economized raw materials and reduced the final filing and finishing operations. The machine pictured here consists of a giant hammer weighing approximately 100 pounds that is dropped onto the base below. The appropriate dies are fixed to the hammer and to the base, and the sudden drop of the hammer stamps the heated metal with the necessary form. The worker on the right fixes the heated metal into the lower die. The two workers on the left raise and lower the hammer with a rope and pulley. The lock shop at Tula contained four such die-forging machines. As the illustration shows, the operation was labor intensive, requiring two forgers and up to three workers to raise and lower the hammer. Drawing from Gamel's description of the Tula Armory, in *Opisanie Tul'skogo oruzheinogo zavoda v istoricheskom i tekhnicheskom otnoshenii* (Moscow, 1826), plate 27, courtesy of the Library of Congress.

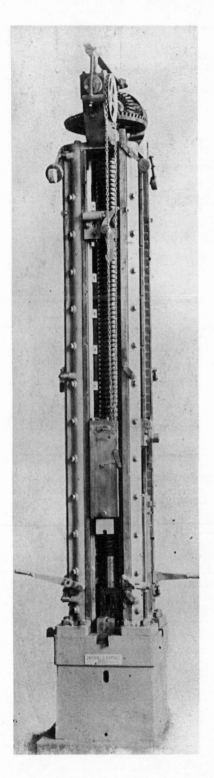

Screw drop hammer with four hammers used in the production of the Berdan 2 infantry rifle at the Tula Armory. Russia purchased this machine from the English engineering firm of Greenwood and Batley in 1871 and put it into operation at Tula two years later. Elisha Root, superintendent at the Colt factory in the 1860s, is credited with inventing the screw drop hammer that forged lock parts using molds, or dies, to determine the shape. This machine, six to ten feet tall, which could accurately shape metal with one blow and required attendance by only one or two workers, represented a considerable advance over the die-forging machine in operation at the Tula Armory in the 1820s. With four hammers, this machine could perform four forging operations successively. Photograph courtesy of the West Yorkshire Archive Service, Leeds.

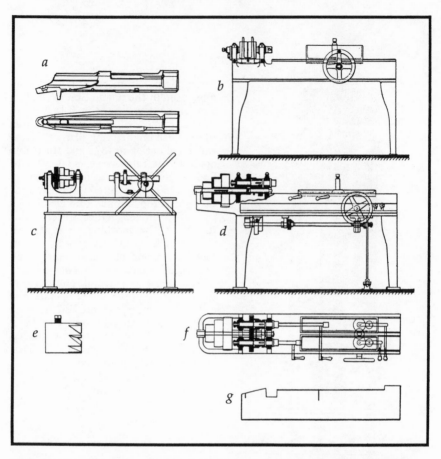

This drawing illustrates the boring machines and lathes used at the Tula Armory to make receivers for the Berdan 2 infantry rifle. Greenwood and Batley sold these machines to Russia in 1871, and they were put into operation in the lock shop at Tula in 1873. The receiver (*a*), a frame at the breech end of the barrel, holds the bolt in line and attaches the barrel to the gunstock. The horizontal boring machine (*b*), shown in the photograph, was used for the first, or rough, boring operations. The lock shop had seven horizontal boring machines, each of which, tended by one skilled worker, could bore up to fifty receivers in a ten-hour working day. A similar boring machine (*d, f - top view*) was used for finer operations. The lock shop had five of these machines, each attended by one skilled worker, that could bore sixty receivers in one day. The boring finishing machine (*c*) removed irregularities in the bore. Two of these finishing machines at Tula processed three hundred barrels each day. The cutters and pattern gauges (*e, g*) were used to make the receivers. Illustration from an article describing mechanized rifle production at the Tula Armory, in *Oruzheinyi sbornik* 4 (1875), courtesy of the Library of Congress.

Horizontal boring machine, used for the first, or rough, boring operations. This machine, purchased from Greenwood and Batley in 1871, was put into operation at Tula in 1873. Photograph courtesy of the West Yorkshire Archive Service, Leeds.

American pattern self-acting rifling machine, used to rifle barrels for the Berdan 2 infantry rifle at the Tula Armory. Purchased from Greenwood and Batley in 1871, the machine was put into operation at Tula in 1873. Rifling was the final mechanized operation on the gun barrel; fine work was crucial for the sake of the ballistic qualities of the rifle and also so as not to waste the previous work on the barrel. An index mechanism on the left side of the rifling machine automatically set the cutter that made the grooves of the barrel. Not needing to set the cutter manually, one worker could tend four machines, and each machine could rifle ten barrels in ten hours with virtually no defects. The barrel shop at Tula also used Enfield and Belgian pattern rifling machines, which could rifle an equal number of barrels but required one worker for every machine. Photograph courtesy of the West Yorkshire Archive Service, Leeds.

remained. As was reported by the factory administration, "Experiencing extreme personal deprivation, many gunmakers have rushed off to find work elsewhere."[5] It is true, of course, that a reduction of state orders on the eve of the tremendous rearmament program had created idle factory hands. But it appears that the earlier fears of conservatives that the state would lose a labor force in slack years were not unfounded.

From the point of view of economic productivity and innovation, the crucial question was *which* gunmakers left the armory—the most successful and most skilled or the poorest and least skilled? Although data to answer this question are lacking, two inferences may be made. Glebov believed that both the most successful and the least successful gunmakers were most likely to leave the armory first. He argued that such a pattern would not hurt the factory: the most successful gunmakers ("armorer-capitalists," he labeled them) had become middlemen and no longer practiced the gun trade; and the poorest of the gunmakers had few abilities anyway.[6] However, considering the different policies regarding senior gunmakers, it might be inferred that, while Tula's senior gunmakers were attracted to stay, those of Sestroretsk and Izhevsk were given the opportunity to leave. Such a pattern would have important implications for future changes in production. A labor force with less seniority would provide less resistance to the imposition of factory discipline, the elimination of home production, and measures to centralize work. A labor force with greater seniority, such as that at the Tula factory, in contrast, would be likely to resist efforts to centralize work, to impose factory discipline, and to curb home production. In fact, as will be shown below, such efforts to rationalize production encountered the greatest difficulty at Tula.

In the Russian countryside emancipation was met with considerable unrest, somewhat unexpected to those authorities who believed that emancipation granted from above would stave off emancipation demanded from below. Confusion regarding the emancipation proclamation in a period during which ex-serfs were still temporarily obligated to ex-landlords fed unrest in the spring of 1861. Similar confusion led to dissatisfaction, if not actual unrest, at the Tula armory. The resulting protest, although it misfired, illustrates grievances during this period, provides a window into administrative practice and the social divisions at the armory, and demonstrates the function of public discussion, or *glasnost'*, the term used then.

In the summer of 1862 a special commission began to prepare the charters for all workers at the armory. However, not wanting suddenly to deprive the factory of its labor force and not having a clear idea of an alternative administrative structure, the commission deliberated for two years. Impatient armorers in protest produced "Voice of the Tula Armorers." As the author(s) of "Voice of the Tula Armorers" stated, the purpose of the protest was to "use publicity [*putem glasnosti*] to obtain an honest investigation of the factory." Originally intended to reach Alexander Herzen's émigré journal, *Kolokol (The Bell)*, the manuscript was seized by the authorities from a Tula shopkeeper; excerpts from it have survived only in testimony presented to the senate. As a result, the protest never became public (*ostalos' bez glasnosti*).[7]

The armorers' grievances essentially were threefold. First, even more degrading then being legally the property of the state, the armorers were in fact the property of the commanders, inspectors, officers, and a myriad of despots. Second, the poorer gunmakers resented their more successful and enterprising brethren:

> The armorers' association is crawling with wealthy masters who have large workshops and who are, in short, lackeys. They swindle the factory's commander, all the while tricking the shop bosses, kowtowing, buying the best jobs, and spying and informing on the miserable. These creatures with all their might fight for and support the existing order at the factory for their own profit, and they get rewarded for their efforts.

Finally, the armorers protested against the corruption and mismanagement of the factory. As the armorers stated in "Voice of the Tula Armorers," "All told at the Tula armory there is a multitude of baseness, violence, flagrant stealing in full view ever since the accession of the anti-Christ [Commander Standershel'd]. This Standershel'd runs the factory as his personal property, squeezing everything out of it he can . . . "[8]

Whether or not this "voice" was the first sign of class consciousness, as a Soviet historian has claimed, the protest did express resentment against the mismanagement of the factory, against the more favored armorers, and against the degradation of the class of gunmakers. The protest would be better described as the first sign of the difficulties facing labor and management in the years following emancipation. But the most important short-term consequence of the emancipation of government armorers was not labor unrest but the perception that management at state armories could no longer remain the same under conditions of free labor. When free labor replaced forced labor, the government could no longer be the paternalistic provider for its labor force; indeed, according to P. A. Zaionchkovskii, the historian of the military reforms, the Artillery Department considered free labor at state armories to be contradictory.[9] The emancipation of the labor force at the armories was part of a complex process that likewise included changes in management, technology, and production organization.

Armory Management

During the era of the Great Reforms, the Russian government relinquished its control over many areas of life in an effort to revitalize and to strengthen the nation and the state. Given the recurring criticisms regarding state management, the time appeared ripe for the government to relinquish management of the armories. This posed several problems. To whom could management be transferred? Would the state relinquish ownership as well? What, if anything, would change in the system of arms procurement, raw materials supplies, organization of production, and labor-management relations?

In trying to answer these questions, the government operated from one basic premise: the need to supply large numbers of arms to the army at the least possible cost under conditions of rapidly changing weapons systems and of freed, and therefore more costly, labor. It was clear that the small arms industry had to attract private capital. At the same time, the Ministry of War, fearful of being completely dependent on private manufacturers, did not want to surrender the armories to the disposal, let alone the ownership, of private persons, whether Russian or foreign. Therefore the ministry offered to lease the state armories to their commanders. The major features of the leasehold management were as follows. The leaseholder did not own the factories; the armories remained the property of the Ministry of War. The leaseholder, who remained an officer in government service, was obliged to provide a certain annual quota of weapons and spare parts at fixed prices to the Artillery Department. The latter body continued to dictate the quantity and types of weapons produced. In this way, the leaseholder acted as a contractor, and weapons production and procurement remained a government monopoly. The government provided the leaseholder with various subsidies: metal was provided free and fixed sums were provided annually to cover construction, maintenance, provision and repair of equipment, light, heat, fire protection, and wages for unskilled workers.[10]

The government assumed that, despite a certain degree of dependence on the state, provision of perquisites and the lure of profits would encourage successful mangement and a mutually beneficial relationship between state and leaseholder. Foremost among the inducements was the opportunity for profit. Although weapons prices were fixed at levels slightly below the previous government purchasing prices, costs were controlled by the manager. Subsidies provided by the state resulted in considerable cost savings. The manager also had full latitude to organize production, to innovate, and to pay workers. With emancipation, management no longer had the burden of providing for a surplus number of workers. Since both the number of weapons and the prices for them were fixed, reduction of unit costs in a large series of weapons would not only lower actual costs but also supplement profits.

Did this hybrid—neither private nor state—system of management, which Zaionchkovskii has laconically described as "strange," work? The first leasing contract was signed in 1863 by Major-General Standershel'd, commander of the Tula armory. This was soon followed by lease contracts with O. F. Lilienfel'd, commander of the Sestroretsk armory, and with P. A. Bil'derling, commander of the Izhevsk armory. The terms of the lease, binding for a period of five years in each case, varied slightly with respect to quantities and types of weapons, prices, and perquisites. It appears that at Izhevsk and Sestroretsk the system worked resonably well. At both factories, the initial lease was renewed, in Lilienfel'd's case for ten years. When in 1884 Sestroretsk finally reverted back to state management, the official history of the armory, generally unsympathetic to capitalism, concluded that the original purpose of the leasehold had been achieved: the armory was supplying one hundred thousand arms to the state, the work was organized in a capitalist manner, the factory had become profitable, and the state had achieved this at minimal costs.[11] However, it must be pointed out that production at Sestroretsk and Izhevsk was already centralized. At Tula, where the organization of production had always posed the greatest problems for cost efficiencies, the leasehold system failed. An examination of that failure reveals much about the difficulties posed by Russian armory practice at this critical time. Beyond this, the Tula experience and the debate it generated in Russian journals reveals much about the attitude toward private industry in Russia.

In Europe private companies were increasingly winning government defense contracts and supplying weapons to the state. Such a pattern was soon to appear in naval procurements in Russia as well.[12] This was an opportunity for Russian private manufacturers to step in to fill an anticipated vacuum in the small arms industry. One of the earliest proposals to take over control of the armories came from the Association of Tula Gunmakers. In 1865 the association, resentful of the leasehold management policy, petitioned the Ministry of War to transfer management of the armory to the association. Not only would the armory be better managed, the association claimed, but the association would also look beyond fulfilling quotas to the state and promote the gunmaking craft in Russia:

> What will happen to the factory and to the gunmaking craft in Russia, and likewise to the gunmakers, when the term of the lease expires of course is of no concern to [the lessor]. If this is the case, one might think that after awhile the number of Tula gunmakers . . . will drop, and that the gun craft at the Tula factory will gradually decline. Then, when it will be necessary to restore [the gun craft], it will require great effort from the government and great expenditures from the treasury.[13]

Such a form of "workers' control" was premature for the Ministry of War and, as far as the sources indicate, never seriously considered. But private ownership was considered and debated on the pages of military journals and at the meetings of the Russian Society of Engineers. Certain advantages of private over government ownership were commonly acknowledged. In his 1862 report, Miliutin complained that at the government armories bookkeeping and paperwork overloaded the directors and kept them from being involved in engineering and production problems. According to V. L. Chebyshev, editor of *Oruzheinyi Sbornik*, private arms manufacturers acted more quickly, which was important at a time when weapons systems were changing rapidly. F. Graf and P. M. Maikov added that government armories were less efficient and produced a costlier arm. Weapons made at private armories, where the owner was impelled by the profit motive, were of higher quality; in particular, private owners decreased the number of defective guns, always the bane of Russian armories, because defective products reduced profits. Since at Tula gun parts were largely made in subcontracting shops, there was no reason final assembly could not also be done in private factories. Finally, argued Maikov, perhaps the strongest defender of private industry, the state could not produce everything; state products would always be inferior, and sooner or later weapons would have to be produced in the private sector.[14]

By and large, defenders of the government armories were unable to come up with strong arguments in favor of state enterprise on its own merits. Chebyshev, not altogether consistently, claimed that a government armory could adopt changes in production more easily. M. Subbotkin, a government inspector at Izhevsk, stated that it was "easier and more natural" for the government to take responsibility for reorganizing the Izhevsk factory.[15] In neither case was it clear why state factories were inherently better than private ones.

The strongest arguments for state enterprise were those against private enterprise. To begin with, Russia had few private manufacturers. This was felt most pointedly when Russian engineering and machine industries were compared with those of Europe. V. N. Zagoskin, writing about metallic cartridge manufacture, noted that France had several private establishments, while "here in Russia, where the machine industry is far behind that of France, it is unlikely that *at the critical moment* we will be able to find private persons who will take on the task of producing armaments."[16] This argument was echoed by the head of a British military delegation visiting in 1867:

> At the present time, when a new arm must be introduced . . . deficiencies in the military establishments of Russia are most apparent. It must be also borne in mind that the almost total absence of private manufactories, capable of giving assistance in time of emergency, throws a double responsibility upon the government.[17]

Moreover, a government that desired to spend as little money as possible despite the needs of a huge army was reluctant to pay the high prices charged by the small number of private establishments. Out of the government's inability or unwillingness to pay came an "anti-kulak" type argument against private enterprise in the arms industry. Critics of private enterprise who believed that private industry would not supply the government at the critical moment really meant that the former would not supply the government at the government's fixed price.

The special circumstances of weapons orders makes the relationship between government consumers and private suppliers unstable. Long periods of small orders, or no orders at all, are interspersed with periods of urgent need. S. A. Zybin, a mechanical engineer at the Tula armory, observed that it would not be advantageous for the government to order rifles or even individual parts, such as barrels, from private suppliers because the private factories would have to shut down production during slack periods. This would cause unemployment and social instability. "The erratic demand for armaments presents an unfavorable condition for their production at private plants."[18] Given the labor unrest at private defense plants at the beginning of the twentieth century, Zybin's words touched upon a serious problem.

Private factories allegedly not able to handle the specialized needs of modern small arms exacerbated the inappropriateness of government contracts with private arms makers. When metallic cartridges and breechloaders were first introduced, private industry, lacking the plant capacity, could not quickly fill large orders; the government had no choice but to turn to the state armories. Private factories were well equipped, Zagoskin argued, to supply copper wire, a semifinished product, but not cartridges and rifles: "The more specialized is the manufacture of an article, the more difficult it is for private industry to fill orders from the Ministry of War, and the harder it is for the latter to count on private industry for assistance."[19] The government paternalism implied in this statement—that private industry would "assist" the state—extended to the attitude toward labor in private industry. Chebyshev was typical when he claimed that private owners in the arms industry would exploit workers and force them out of the craft, leaving the government without gunmakers. It was also commonly argued that despite the lure of profits private owners would have no incentive to incur the expense, and risk, of plant modernization after they had concluded a contract for deliveries to the state.[20] It is clear that the proponents of government enterprise hardly expected private owners to behave as rational innovators and profit maximizers.

The debate over private and state control of the small arms industry reveals important features of Russian industrialization and entrepreneurship. The arguments against private ownership in the small arms industry had, at their roots, two premises. The first premise, stated simply, was that there were not enough private factories and that therefore the government had no choice but to turn to state industries. A variation of this premise was that the technical capabilities and plant capacity in the private sector were inadequate to do the job. In essence, this was Zagoskin's criticism of the munitions industry of the late 1860s when Russia was rapidly adopting metallic cartridges. However, writing in 1875, Zagoskin noted that because of government assistance the technical capabilities of private factories had improved. This branch of the armaments industry during the 1860s and the 1870s provides an early example of Witte-style pump priming of the private sector.[21] If one's arguments against private ownership in the arms industry were based on the premise that there were too few private arms factories, then expansion of the private sector would have allowed the government to turn more and more to private manufacturers, as did the Naval Ministry.

However, as the inventor and engineer N. V. Kalakutskii argued before the Russian Society of Engineers, the number of factories, technical sophistication, or plant capacity in the private sector were irrelevent: the needs of the army for small arms must be met exclusively by state factories.[22] This argument was based on a second premise, the fundamental mistrust of private manufacturers and a fear that the government would become dependent on private producers who, "at the critical moment," would hold the state hostage. Unlike the provision of battleships, for example, provision of small arms was a vital link in the nation's security. The root of this fear of dependence on private producers—the mistrust of private gain and the alleged rapaciousness of the individual entrepreneur in an economy of scarcities—went deep in Russian culture and was shared by government and people alike.

The arguments against private ownership became even more persuasive when coupled with the dissatisfaction about the day-to-day operation of the leasehold armories, particularly at Tula. In outlining anticipated management changes in 1862, Minister of War Miliutin had called attention to the need to make arms quickly and more cheaply using modern machinery. However, the management changes did not appear to have the desired effect, and in 1868 two artillery officers, Generals Glinka and Notbek, were dispatched to the armories to investigate delays in deliveries, difficulties in changing production to Krnka conversion models, and the age-old problem of defective weapons. One year later Miliutin himself visited the flagship Tula armory and was disappointed that so little progress had been made.[23] The leaseholders had little incentive to innovate and, as a result, were having difficulties producing breechloaders. At Izhevsk, already fairly centralized, the lessor had little incentive to mechanize, and hand labor was still widely used. Even more discouraging was the practice at Tula where neither mechanization nor centralization was adopted; General Standershel'd, the lessor, had preserved the system of subcontracting work outside the armory. Just as prereform agricultural inefficiencies were preserved after the emancipation of the serfs, so, too, were inefficiencies in labor and management preserved after the emancipation of the gunmakers. With large orders of the Berdan rifle anticipated, Miliutin concluded that leasing the Tula armory was no longer advantageous to the state. When Standershel'd's lease expired in 1870, Tula reverted to government control. Miliutin appointed General Notbek to be the new commander at Tula and to go abroad to study foreign armory practices.[24]

Thus, the Russian government admitted that the leasehold management system—at Tula at any rate—had not performed its desired function to mobilize private capital and entrepreneurship for the purposes of innovation, mechanization, and reorganization of production.[25] Miliutin's decision to nationalize the Tula armory had important undercurrents. First, Notbek observed that "unimproved machines did not fit the job and could be used only with considerable hand finishing of parts."[26] This confirms the earlier conclusion that the "superb" machines introduced a half century earlier had not been improved or integrated into the production process. Second, the need to reorganize the armory and adapt it to "rational production," in Miliutin's words, was perceived not when earlier breechloaders such as the Carl and Krnka were adopted but precisely at the moment when the Berdan model had been selected for domestic production. Although the Berdan was preferred for its simplicity, its fabrication nevertheless posed a sufficient challenge that forced a reponse in Russian industrial organization and armory practice. Finally, it is significant, given the centralization of production already attained at Sestroretsk, that Notbek should have been ordered to seek models of rational production from foreign armory practice. Rationalized, centralized production suggested above all the form of armory pratice employed in New England. Indeed, Russian artillery officers were already in New England, not only to design, to test, and to procure weapons but also to observe American armory practice.[27]

Russia and the "American System"

Although no royal commission or parliamentary select committee from St. Petersburg visited the United States, news of mechanization, centralization, and interchangeability spread to Russia. Descriptions of the Enfield armory provided initial exposure. In 1861 the specialized Russian press observed that Enfield was the only European armory using "superior machines from the United States." A report of the Paris Exposition of 1867 likewise noted that at the Enfield armory, considered among the best in Europe, all work was done on American machines. The resulting division of labor had progressed so rapidly that it took ten times as many separate operations to make an Enfield rifle than to make the Russian .60 caliber musket.[28] In several articles the editor of *Oruzheinyi Sbornik*, V. Chebyshev, claimed that by the 1850s the Americans had apparently solved the problem of machine production "brilliantly." Chebyshev unequivocally defined American armory practice as the operation of sequential machines and, insofar as possible, the replacement of men by machines. "Presently, the so-called American, or mechanized, method of manufacturing is getting serious attention. As we know, the essence of this method is the widespread application of milling cutters and milling machines to the production of firearms. This permits not only preliminary working of the metal components but also working them to a nearly finished state. . . . One frequently hears that to improve work at armories all you have to do is convert to machine production and order machines from America."[29]

Although the American system seemed promising, Chebyshev expressed five reservations. First, the American system had been adopted only at Enfield and to a certain degree at Sestroretsk, even though it had been well known for more than a decade. This fact alone suggested that American armory practice might not be appropriate for all situations. Second, machines did not necessarily speed up production. To prove this point, Chebyshev noted that even "the best American factory [Colt]," when approached by Russian agents, required nine months to prepare the machines and tools and that no more than one hundred arms per day could be made. Third, interchangeability did not necessarily result in exact fitting; consequently, some operations were better done by hand. This argument was voiced in America and England one generation earlier, particularly by those who remained unconvinced of claims by manufacturers such as Colt. Fourth, factory discipline did not always yield the best results; the hard-working craftsman, so the common argument ran, could accomplish more at home, especially with the (free) help of members of his family. Finally, the slightest change in specifications necessitated great expenditures; therefore the higher cost of arms made with special purpose machines offset the many advantages of a machine-made product.[30]

Chebyshev was not merely expressing the reservations of a traditionalist concerning new production methods. For a poor country modernizing its weapons, the cost factor was critical. And the cost factor was most critical at the stage of design testing and modification, at the stage today known as research and development. If, as both Howard and Hounshell argue, mechanization initially raised rather than lowered production costs, and if high costs prevented American private manufacturers such as Colt from adopting complete interchangeability, one can easily imagine the obstacles machine production presented to the Russians.[31] In fact, Gorlov and Gunius, visiting the United States to select the best rifle and metallic cartridge system, were at first confounded, not aided by the American system. Not having a completed design that could be licensed to a manufacturer for production, the Russian officers found it difficult to negotiate a potentially costly production contract. Given that American manufacturers had introduced costly special-purpose machines requiring a series of jigs and fixtures,

> every change in the model requires the installation of new machines and with it costs which the factory may absorb only if it can count on recompense with a large order in the end. Since [our] agents did not yet have a final model but, on the contrary, anticipated a large number of changes, they considered it risky to enter into a contract with a factory.[32]

Yet the Russians had no choice. Despite the obstacles, the United States provided the crucial research and design laboratory—the facilities for testing weapons designs, machines, and production—that Russia could not provide for itself.

Despite Chebyshev's reservations about mechanization, the Russians in the end realized that machinery and interchangeability were the methods and goals of future small arms production. It is worthwhile to keep the fundamental rationale for interchangeability—battlefield performance—in mind. The Russians were quite practical in this regard and were not looking for perfectly made, finished, and filed weapons. Although not a panacea, interchangeability was no less important. Buniakovskii stated the Russian position well:

> Despite all the improvements it is obvious that it will be impossible to attain the kind of firearm in which all parts would be of uniform strength. The principle on which the Russian arms made in America is based, that is to say, interchangeability, will make it easier for the soldiers to repair this arm than to repair the muzzleloader without a machine or a gunsmith.[33]

Buniakovskii grasped well that interchangeability was not an absolute but a matter of degree. He acknowledged that a uniform arm would be impossible to produce and concluded that the Berdan, production of which he was inspecting, would not offer perfect interchangeability in the field, although it would be easier to repair than previous models.

Since Hiram Berdan, the designer of the rifle and metallic cartridge system adopted and modified by Gorlov and Gunius, did not have a manufacturing plant, the Russians were forced to turn to a manufacturer. That Colt eventually received the Russian order for thirty thousand Berdan infantry rifles in 1868 was due in part to the company's perception of an excellent business opportunity and to earlier contracts with the Russian government. Two additional factors may be added here. First, of course, was the reputation of the Colt factory for centralized mass production using sequential special-purpose machines and precision tools to achieve interchangeability. Second, and perhaps an even more important factor, was the concurrence of the Colt practice of inside contracting and the Russians' needs in system testing and alteration and in mastering the art of centralized mass production. William Franklin had offered space in the Colt factory so that Gorlov and Gunius could conduct their tests on rifle and cartridge systems. Although the Russian officers had to pay for the labor and for the changes in machines and tools, the Colt facilities were available as a research and design laboratory for system design and testing and for hands-on experience in American armory practice. This experience buttressed several arguments in favor of mechanization at a time when it was being intensely debated. These arguments can be found in the reports of Buniakovskii, the officer sent in 1869 to Hartford to supervise production of the Berdan rifle.

Buniakovskii admitted that production of the Berdan entailed many difficulties. First, despite the Berdan's relative simplicity compared to other systems, the mechanism was "not suitable for production by hand and was difficult even for machine production." In some ways, of course, this was a blessing in disguise for those who wanted to abandon hand production. Second, the small-caliber barrel was difficult to mill and required the highest quality steel. American steel for gun barrels was considered to be of lower quality than European steel, and this had posed a perennial problem for Colt. Although the Obukhov steel mill had recently started supplying the navy, the possibility of dependence on foreign suppliers of steel posed an obstacle to a completely domesticated arms production.[34]

According to Buniakovskii, Colt had solved the various production difficulties. If Russia were eventually to commence production of the Berdan rifle, Colt's experience suggested to Buniakovskii four lessons. The four lessons provide an insight into the process of technological transfer and illustrate a progression of complexity of the borrowing process from simple equipment purchase to adoption of an entire culture of work. First, Buniakovskii recommended purchase and replication of machinery: "It would indeed be helpful to not lose time and order here . . . American vertical boring machines and lathes. . . . It would not be especially difficult to make an exact copy, not only of all the machines but also of the tools, used to produce the .42 caliber rifle." Second, Buniakovskii suggested that *before* the machines and tools were shipped to Russia they be installed at the Colt factory itself for Russian use. Here, the Russians could use the inside contracting practices as a laboratory in which to master production processes. Buniakovskii noted that twenty-two separate contractors were working on the Russian order. The third lesson provides a textbook case of the "advantages of backwardness," whereby the latecomer can skip stages in the development of industry. Buniakovskii advised the Ministry of War to use the Colt machines to

> assemble a ready-made mechanized armory. . . . Otherwise, to set up the production of smaller-caliber barrels in Russia would take up considerable time and much trial and error, which the [Colt] factory has already gone through. . . . If it is decided to build in Russia a factory for precision machine production of small arms, but of another system different from that being currently made in America, then it will be much more difficult and take much more time than making a copy from an operating factory.[35]

Backwardness has its disadvantages, and it was the fourth lesson that had the most far-reaching consequences for the culture of Russian industrial production. Buniakovskii saw beyond the mere installation of American machines on Russian soil: "We have a very convenient opportunity to apply precision machinery and work to Russian armories." To make his case, Buniakovskii began by stating that what was most instructive about the Colt factory was its "practical mechanics and unique organization." Like all American institutions but unlike all Russian institutions—as it no doubt seemed to Buniakovskii—the Colt factory had a very small administration. Moreover, there were more machines than workers. Mechanization and the high degree of interchangeability meant that final finishing and assembly required only four workers at Colt but hundreds of workers in Russia. The workers at Colt were "accurate and clean," and the mechanics "got their reputation from their practical, not their theoretical, knowledge." Russian industry could advance not only by copying machines and tools from the Colt factory but also by recruiting skilled workmen. Buniakovskii proposed recruiting "the most skilled workmen, who in two to three years would transfer their skill to Russian workmen." This transfer of skills, according to Buniakovskii, would educate native workmen, improve the quality of engineering in Russia, and raise wages and thereby the standard of living of gunmakers. (At least one Colt inspector was sent to Russia, although company records do not indicate the length of time he spent there or his responsibilities.) Reproducing the Colt factory in Russia, bringing Colt mechanics to work in Russian armories—in short, nothing less than "establishing in Russia a mechanized armory on the American model"—would provide a great opportunity to revolutionize the Russian small arms industry.[36]

Thus, as the Colt facilities allowed American independent subcontractors to improve system design and production processes, so too they provided the Russians an important laboratory at the moment when Russian armory practice, particularly at Tula, was being reexamined. The emphasis in New England armory practice on process over product provided the learning model that had so far eluded the Russians in their efforts to rationalize small arms production. Although Russian officers, being military men, not social scientists, did not provide a label, they noted the process orientation of other American manufacturers as well. Marveling at the productivity of the Providence Tool Company as it filled orders of the Peabody-Martini rifle for the Turkish government in 1877, A. Fon-der-Khoven observed that the factory was able to change manufacturing quickly from articles for the shipping industry to firearms when the Civil War broke out and that part of the work was contracted out to skilled specialists.

Fon-der-Khoven commented that although this might seem strange, "it is nothing to an American. . . . This method is widely used in America at many factories."[37]

The Russian officers in the United States frequently commented, usually approvingly, on the American penchant for machines. Both Buniakovskii and Bil'derling noted the large number of machines relative to the number of workers, the adaptation of the machine to the production process, the reputation of a mechanic based on practical, rather than theoretical, knowledge, and the precision and orderliness of tools and workmen alike. V. Eksten put it somewhat more bluntly: "Scornful of American hand labor for its imprecision and above all for its high cost, that country's factory owners make firearms almost completely by machines and substitute milling machines for all the work done elsewhere by skilled workmen." Bil'derling also noted the diligence of the American worker, no doubt unusual when compared to that of the Russian. "The American cannot stand idle at his machine; he adjusts the machine to operate on its own and while the machine is operating, himself does another task."[38]

The American penchant for machines and the results it produced were even more striking when compared to the nature of work at the Birmingham Small Arms Company, where the Russians had also contracted for an order of Berdan rifles. Captain Bil'derling, inspector of the small arms made for Russia in Birmingham, reported a great degree of dissatisfaction in the work at Birmingham to the Artillery Department:

> The machine production I saw in America is a long way from acceptance here in England where hand production reigns supreme. The factory here is just as unfamiliar with the small-caliber rifle as we are. They admit that they have never made small-caliber arms and consequently had to learn on our order. Having learned at our expense, at the end of the order they will cease being accommodating and will require us to pay a huge amount for the patterns, tools, and machines, which in fact we specified. As for the quality of English workmanship, it is much lower than American. The shocking percentage of defects in the parts has barely decreased: 50 per cent of the barrels and stocks, 30 per cent of the frames, 25 per cent of the mechanisms, and 15 per cent of the ramrods are defective.[39]

The production experience gained at Colt was underscored by comparison with the absence of such experience at Birmingham.

The advantages to Russia of building a mechanized armory on the American model went beyond the construction of any single rifle system. The superiority of the Berdan 2 over the Berdan 1 "was not so much in its military capabilities as its promised superiority for large-scale production at our armories."[40] Because of its order of small arms in

America, the Russian government believed that it had an opportunity to surpass Europe in small arms technology; that is, to take advantage of backwardness. Large initial costs would be redeemed by "spectacular results." As Miasoedov later recalled,

> We borrowed the machine method of making firearms chiefly from America, where interchangeable wares are made with such precision that one by one our officers, seeing this production for the first time, not only brought back their full approval, but also the conviction that such precise work was still unthinkable here at home.[41]

Although the cost of advanced machinery and foreign engineers would be high, the cost in training and paying skilled craftsmen rather than mechanizing production in a poor country such as Russia was likely to be even higher. Berdan himself pointed out to the new commander of the Tula Small Arms Factory, General Notbek, that "the way he was going now he could never make interchangeable work in Tula without filling the Armory with first class tool makers from other countries."[42] Later, Miasoedov described the revelation as the new concept in the organization of work became apparent:

> At first it was incomprehensible that America did not value skill for its own sake; that machine-made interchangeable parts were not made by a caste of gunsmiths, as in Russia, but often by women and even children; that the machine makes it possible to do without a caste of gunsmiths and to have a relatively small number of workmen trained not in a craft but in an understanding of the processes and work of machines, in precision measurement, in the assignment of work, etc.; and that other workmen are merely operatives, making machine production advantageous precisely where there is not a large number of skilled workers. Little by little, these things became so clear to us that we ourselves decided to redesign our small arms and ammunition plants for machine work.[43]

The wonders promised by the American system were, of course, based directly or indirectly on the observations of Russian officers in the United States. Gorlov, Buniakovskii, Bil'derling, and Miasoedov witnessed the machine in the American garden. But the ultimate goal was to implant on Russian soil the mechanized, centralized American armory practice. The machine had to grow in the Russian garden. The reorganization of the Tula armory provides insights into this process.

The American System in the Russian Garden

"The Americans have not only sent arms and ammunition in large quantities, but it appears they have done Russia the much greater service of supplying her with the machinery which will enable her to make her own guns, and so be independent of external help."[44]

Mechanization and Reorganization of Work at Tula

The rearmament of the Russian army in the late 1860s and the 1870s affected the organization of production at all three state armories. Russian artillery officers concluded that the .42 caliber Berdan rifle could be made only by machinery. Although the Tula armory had the largest order for Berdan rifles, much work was still done by hand in more than two hundred private workshops, and the factory buildings were inadequate to meet the needs of a major contract for machine-made weapons. Beginning in 1868, the three government armories were ordered to begin mechanized production "according to the model of the most famous American and English factories."[45] Consequently, in 1869 the Ministry of War formed the Commission to Reorganize the Armory, headed by General of Infantry Vladimir Vasilevich von Notbek (1825–1894), the armory's new commanding officer. The commission included an ordnance expert and Notbek's assistant at Tula, also an editor of *Oruzheinyi Sbornik*, General of Artillery Vasilii Nikolaevich Bestuzhev-Riumin (1835–1910), and the future Minister of Finance I. A. Vyshnegradskii (1831–1895), a professor of mechanical engineering at the St. Petersburg Institute of Technology.[46] Their report recommended changes that went far beyond the anticipated change in administration. The most important recommendations may be grouped around two broad goals—mass production of firearms and the use of standardized components—and two broad means—mechanization and centralization.

The Notbek commission recommended that the armory be mechanized to produce annually seventy-five thousand Berdan rifles with interchangeable parts. The government's decision to change the system of making firearms at state factories from handicraft production to machine production served notice to the armory superintendents that interchangeability was to be the goal of the production process.[47] In a sense, this decision was akin to that of the American Ordnance Department when it granted a contract to Simeon North in 1813 for the production of rifles, with the stipulation that the parts be interchangeable. However, the Russian government did not have the time to develop leisurely a native machine-tool industry. The absence of private manufacturers able to take a large order for machines forced the government to turn to foreign suppliers. To outfit the Tula armory alone required hundreds of machines, which the government tried to purchase at the lowest possible cost. Although Hiram Berdan lobbied in St. Petersburg on behalf of the Colt Company, Colt prices were too high, and the first contract for machinery was awarded to the Birmingham Arms Company in 1870. Finally, in April 1871, the largest order of machines (854) was purchased from the English engineering firm of Greenwood and Batley in Leeds at a cost of one and one-half million rubles.[48]

In 1872 Greenwood and Batley dispatched an American, James Henry Burton, and a team of English workers, as Thomas Greenwood put it, "to assist us with your experience in cutters, gauges and other necessary things to set the works going and to go out to Tula and superintend the starting of the plant."[49] Burton himself was a key link in the transfer of the American system to Russian armories. Burton had been a master armorer and machine-tool builder at Harpers Ferry. He helped transfer the American system to the Enfield armory and later worked at the Birmingham Small Arms Company before going to Greenwood and Batley. Under Burton's supervision, hand labor was eliminated, boring, cutting, and milling operations were mechanized, and special-purpose machines were put into sequential operation. Moreover, the installation of three 360-horsepower turbines enabled the factory to operate year around, even during the spring flood of the Upa River. The factory reopened in 1872 and one year later regular production commenced.[50]

Centralization of production involved two alterations. First, the existing armory was rebuilt and reorganized. After 1872 the factory buildings formed a closed rectangle surrounding a foundry shop with fifty ovens, fourteen presses, and two steam hammers. The factory was redivided into ten shops, the new ones being the tool, fitting, and conversion shops, and a model machine shop. This rationalization of production organization caused the armory to conform to special-purpose machines, made the shops convenient for the transfer of work and for supervision, created a self-contained factory, and, finally, provided for the repair of the machines themselves.[51]

Second, and arguably more important, the subcontracting of work at the premises of the gunmakers was curtailed, and all jobs were henceforth performed on factory grounds. The gunmakers were now subject to the strict supervision of the factory administration for the duration of a ten-hour day. Moreover, despite persistent concerns of the government, mechanization drove some gunmakers out of work. During the first year, the workforce declined from three thousand to fifteen hundred; according to the Soviet historian Ashurkov, many of the gunmakers found new jobs in the employ of local mechanics. For those who remained at the factory, wages jumped by 30 to 40 percent.

The prerevolutionary military encyclopedia gave much credit to Notbek himself: "In six years Notbek completely retooled the factory to make the small-caliber rifle highly efficiently, while putting the factory significantly ahead of famous foreign armories in a technical sense."[52]

After 1873, the Berdan 2 became the major product of the Tula armory. By 1884, when rearmament with the Berdan 2 was completed, approximately one million Berdans had been made at Tula. Ironically, Russia had barely finished manufacturing the Berdan rifle when European armies began to adopt even smaller caliber magazine rifles. At the height of the Turkish War in 1878, Tula made 154,000 Berdan rifles at a rate of approximately 650 per day.[53] In addition, the third model Smith and Wesson revolver, adopted by the Russian government in 1871, was also made at Tula; 300,000 to 400,000 such revolvers were made during the 1880s and the 1890s.[54] The unit costs of weapons declined by almost 50 percent, and the annual 280,000 ruble savings in wages alone made up the government's original expenditures of 2.9 million rubles in ten years. After its reconstruction Tula was never behind in its weapons deliveries to the government. Staffed by skilled officers from the Artillery Academy, Tula became the technological leader of the government armories. In 1872 the Tula factory sent a set of machines to the Moscow Exposition of Science and Industry. To spread the gospel of armory practice, one year later a permanent museum was opened at Tula, in those days a sure sign of satisfaction in the armory's performance.[55]

Foreigners and Russians both were impressed with the results. According to Frederick Wellesley, British Military Attaché,

> The present works are entirely devoted to the manufacture of Berdan rifles and are hardly yet completed. They cost more than three million rubles, more than half of which sum was spent in the purchase of machinery. The works are immense, well organized and well supplied with machinery purchased for the most part of Messrs. Greenwood and Batley of Leeds.[56]

In the words of the catalog of the 1872 Exposition of Science and Industry, "barrel turning lathes were built for the first time at Tula. . . . Even those at the Colt factory with its high degree of machine production still have not been able to do this job on a self-acting machine."[57] Twenty years later the authors of a study of Russian industries for the Chicago Columbian Exposition in 1893 proudly proclaimed that the Tula factory was "justly considered to be one of the most important and well organized of the European manufactories;

all parts of the gun are manufactured by machinery and the exactitude and finish as well as the adjustments of the several parts are not inferior to those of the best foreign makers."[58] Even Soviet historians have been generous in their praise for the armory's accomplishments. The fact that Tula never fell behind in its arms deliveries to the government allegedly showed that the gunmakers had mastered the new machine technology, though later problems should make one cautious in one's praise. In the opinion of Soviet historian Ashurkov, the armory after reorganization was just as good as those in Europe.[59]

Although Tula was the largest and, after its reorganization in 1872, considered to be the most advanced of the three government armories, small arms in general and the Berdan infantry rifle in particular were machine-made at Sestroretsk and Izhevsk. Each of the other two armories had to a certain degree adopted the centralized factory mode of production earlier than Tula. And Sestroretsk and Izhevsk had some claim to specialization in the technologies of specific gun parts: the former was known for its advances in stockmaking, and the latter improved the technologies for making steel barrels.[60] Finally, metallic cartridges were made at a government cartridge factory in St. Petersburg.

Sestroretsk, Izhevsk, and the Petersburg Cartridge Factory

Situated in close proximity to the birch and pine forests of northwestern Russia, the Sestroretsk Small Arms Factory was seemingly in a good position to respond to the developments in woodworking technology. Unfortunately, its resource endowment did not match the capabilities of early stockmaking machines: birch lent itself less easily to woodworking machines and to inlaying than did walnut. By hand, a stockmaker could fashion one stock per week with great difficulty. When, however, in 1857 Captain Lilienfel'd brought back from America an infantry rifle with a machine-made birch stock, the Committee on Improvements in Small Arms decided to test several American machines at Sestroretsk. Tests showed that stocks could be machine-made from birch as well as from walnut and that a complete stock could be cut in fifteen minutes.[61]

With uncharacteristic speed, in 1858 the Russian government ordered a set of seventeen machines from the Ames Manufacturing Company of Chicopee, Massachusetts, which were put into operation in 1860. This was part of a project to adopt American machine production at Sestroretsk, which also included purchase of stockmaking and rifling machines from Colt. Only six machines needed a separate operator; the remaining machines could run with one operator for every two to four machines. Using the machines, one worker could prepare forty to fifty blocks of wood per day and between two and three completed stocks per day. By 1865 the machines had practically eliminated hand work and had caused daily wages to fall. It was feared that machines would in time cause the art of hand stockmaking to disappear.[62]

A delegation of British artillery officers visiting Sestroretsk in 1867 found that "the shaping and cutting room of the stock department was not much larger than a good sized bedroom. The machines chiefly on the copying principle, are good. They are American and all, except that for boring a hole for the ramrod, exactly similar to those employed at Enfield."[63] According to Evgenii Gutor, author of a description of gunstocking machinery, the American machines were complex and at times more clever than useful; they were not easily integrated into production. Like the special-purpose machines at Colt, they were more process than product oriented: "In executing with extraordinary precision the most difficult woodworking jobs, they unite a large number of the most varied transformations and transfers of movement, a study of which could be indeed beneficial for each mechanic planning new mechanisms."[64] Herein, of course, lay the significance of the Colt machines and of the American system. Individual machines had been installed at Sestroretsk before; however, this was the first time that a series of machines had been brought into production.

Sequential operation of machines hastened the division of labor. S. A. Zybin, a mechanical engineer and author of a popular history of the government armories at the turn of the century, marveled at the "complete division of labor" and continued,

> Work is distributed here such that each worker needs only to grab hold of a piece, turn on a machine, and after the work is finished, turn off the machine and remove the piece. All manipulations are exceedingly simple: after a brief instruction on how to operate the machine, any laborer, straight off the farm can start to work. Thinking and dexterity in this work have almost no place. You need no more than habit and with this the worker, all day tending the same machine, can at times achieve striking productivity.[65]

Like the United States, although for different reasons, Russia too had a shortage of skilled labor, and such extreme division of labor promised an opportunity to recruit and to train an unskilled labor force.

For a long time barrel production had been the most serious bottleneck in the small arms industry. Barrel forging by hand was very slow (one worker could do three per day) and resulted in a high number of defects. The use of steel and the spread of mechanized rifling in Europe in the middle of the nineteenth century made this bottleneck even more acute. Located in close proximity to the iron ore deposits and mining industry of the Urals, the Izhevsk Small Arms Factory should have been in a position to contribute advances in barrel production. Instead, for many years mechanization proceeded in fits and starts, and it was not until the introduction of centralized armory practice in the 1870s that genuine progress was made.

From 1809 to 1866, the factory produced approximately one million muskets, most with iron barrels. Although Russian military agents in the United States reported that high quality barrels had been made with Russian iron, rifles required steel rather than iron, which was a weaker material. Russian arms manufacturers, as had American arms manufacturers until the 1860s and the 1870s, found the quality of domestic steel deficient and turned to foreign suppliers, chiefly the firm of Berger in Westphalia. This, of course, added to the expense of an already expensive product in a poor country; it also made the Russian army dependent on supplies of foreign steel, undesirable in wartime, as weapons experts never tired of pointing out.[66] Given the poor-quality materials, the introduction of machines was not necessarily a rational decision. A double-barreled boring machine and mechanized barrel forging were introduced in the 1840s, but barrels were made primarily by hand through the 1860s. Mechanized barrel forging, introduced in 1856, initially increased the number of defective barrels; machine work was curtailed and hand work was reestablished. The fabrication of locks and stocks posed similar difficulties. The lock stamping and stockmaking machines that "appeared," in the vague phrasing of Aleksandrov, in the late 1850s and early 1860s were not put into production. Since the miltary authorities were interested in cheaper hand labor, argues Aleksandrov, these operations were not mechanized until the late 1870s.[67]

Modernization of production at Izhevsk took three forms. First, with the establishment of the Izhevsk steel mill in 1870, Izhevsk not only ceased importing foreign steel but even began exporting finished steel barrels to other armories.[68] A domestic source of steel made possible the second form of modernization—mechanization. As of June 1872, about 5,000 armory workers were still making Krnka conversion rifles by hand. That mechanization had become a top priority for the government was evident in the terms of the second leasing agreement in 1873, which directed the superintendent of the armory, Petr Aleksandrovich Bil'derling (1844–1900) to mechanize barrel production. As at Tula, the impetus for mechanization came with the need to equip the factory to make the Berdan rifle, and between 1872 and 1873, the factory was refurbished with 486 new machines, many of them American. Rifle production increased to 60,000 annually. In 1875 the factory was equipped to commence production of the Berdan, and three years later more than 53 percent of the workers engaged in the most important steps of gun production operated machines.[69]

The use of steel and mechanization facilitated the third form of modernization—specialization and centralization. However, changes in the organization of work were not as dramatic as those at Tula. The structure of the shops did not change after 1866, the date of the abolition of serfdom at Izhevsk, and some of the workers continued to maintain small workshops at home. Nevertheless, the production process did become more complex, more mechanized, and more dependent on the uniformity and precision dictated by the Berdan model. As at Tula, an increasing number of operations in the production of gun components subdivided shops and diluted skills. By the end of the 1870s, separate boring and turning shops had emerged in the barrel trades, and separate stamping, milling, and fitting shops had emerged in the lock trades.[70]

Greater specialization and mechanization were not without unpleasant side effects, and advances in machine production, as in the United States, were introduced slowly. The frequent model changes, particularly in the late 1860s, caused work stoppages. In 1868, for example, the number of workers employed dropped to 50 percent of the 1864 level. Proper gauges, instructions, and tools did not always arrive at the factory on time. And, despite the increasing sophistication of machinery and production processes, the main power source remained an adjacent stream that was often too weak to turn the wheel, stopping work. Moreover, the introduction of milling machines and lathes for making barrels continued to meet considerable resistance from the military establishment. Chebyshev, in particular, argued that machine-made barrels were still inferior to hand-made barrels. Isolation continued to be a problem for Izhevsk, making it more difficult to become an innovating armory; even after the new lease of 1873 and the ensuing mechanization, Izhevsk continued to borrow its machine technology from the more advanced armory practice at Tula. Nevertheless, eventually one million Berdan rifles were made, and *Russkii Invalid* (Russian invalid), the official army newspaper, was enthusiastic about the changes at Izhevsk:

> The refurbishing of the armory in such a short period and its adaptation to machine production of the very best firearms, along the lines of the most famous American and English plants, undoubtedly counters our predilection to have no faith in our own productive ability.[71]

The choice of rifle conversion system and the eventual adoption of the Berdan was in great part driven by the decision to adopt metallic cartridges. As in the case of rifles, the decision to adopt was taken with the future needs of domestic manufacture in mind. When the Ministry of War borrowed from the United States the technology for mechanized manufacture of rifles, it also made arrangements for domestic production of the appropriate ammunition. The first machines to make metallic cartridges were set up in 1866 in the capsule division of the Okhta ammunition plant in St. Petersburg. One year later the St. Petersburg Cartridge Factory opened; it included two metallic cartridge case shops under the authority of the army's Executive Committee for Rearmament. The cartridge factory was instructed to make one million cartridges annually, to organize mechanized mass production, to acquire experience in the construction of special-purpose machines, and to train skilled workers. At the same time that the Union Metallic Cartridge Company was filling a large Russian order for center-fire metallic cartridges, Gorlov ordered American machines, patterns, tools, brass, and even skilled workers from the United States to equip the St. Petersburg factory. Since Russia did not have the proper working and measuring tools, acquisition of a series of precision gauges, patterns, and cutters was a key step toward the manufacture of small-caliber metallic cartridges.[72]

In 1871 the metallic cartridge case shop was rebuilt for work on ten series of machines imported from the United States or built on American models in Russian factories. At first, the factory made cartridges for both the Krnka conversion rifle and the Berdans, but, after 1873, the entire production line was devoted to cartridges for the Berdans. By 1874, the St. Petersburg Cartridge Factory was producing twenty-five thousand Berdan cartridges per day on twenty-two series of machines.[73] Oddly enough, the greatest production problem the Russian government faced in this case was not with the installation of machinery—many of those production problems had been worked out in the United States—but with a critical raw material, copper, needed for the brass cartridge cases, percussion caps, and bullet casings. According to Charles Norton, the American specialist in manufacturing, Gorlov "freely admitted that the success and adoption of [the Berdan] cartridge by the Russian government was due in large degree to the admirable character of metal supplied by the Coe Brass Company."[74]

Russian government contract policy deviated from that of the American government and thereby prevented more rapid development of the native cartridge industry. In 1873, one General Kartashevskii proposed building a private cartridge factory for the Artillery Administration. Although the Artillery Administration was willing to contract for fifteen million metallic cartridges, it refused to offer any cash advance or subsidies. It was not until ten years later that a private cartridge factory finally opened in Tula.[75]

All evidence suggests that the adoption of small-caliber metallic cartridges was one of the most successful aspects of the Russian program to rearm and to domesticate foreign technology. Russian officers were convinced that Russia had the best product available and that Russia would be better armed than any other country. One officer, expressing regret in an 1880 article that he had been unable to visit American cartridge manufactories, added, "We are quite familiar with their cartridge production, because it was transplanted here and in so doing was altered and improved."[76]

More significantly, the import of precision measurement tools and skilled workers hastened the development of a precision tool industry, so important in the diffusion of technology. Before 1868, when calipers appeared in Russia, even mechanics and opticians could not measure more precisely than 0.06 mm. However, metallic cartridges required a precision of 0.006 mm. P. Kharinskii declared that the American skilled workers who came to the St. Petersburg Cartridge Factory were "our first teachers in cartridge manufacture" and that for ten years the St. Petersburg factory "has been a school where mass cartridge production was undertaken for the first time in Russia."[77] As a result, according to Zagoskin's report to the Russian Society of Engineers, by 1870 the St. Petersburg Cartridge Factory was using Russian-made tools accurate to 0.0001 of an inch; that is, five time the accuracy of tools used in American factories.[78] The study of Russian industries prepared for the Columbian Exposition of 1893 stated, "To what degree of exactitude this industry has attained may be seen in considering the delicate requirements of cartridge making. It is no longer difficult to limit the dimensions of the cartridge to 0.001 of an inch and of instruments to 0.0001 of an inch."[79]

By the early 1870s Russian small arms production finally had been converted to the centralized armory system perfected a generation or two earlier by government armories and private arms manufacturers in New England. Production was centralized in a single factory under strict supervision and control, subcontracting in private homes and workshops of individual gunmakers had been eliminated, production processes were mechanized using the sequential operation of special-purpose machines, and jigs and fixtures permitted a high degree of interchangeability with only a small amount of filing and finishing done by hand. An additional component of the American system was a machine-tool industry capable of manufacturing sophisticated, special-purpose machines. The technological convergence provided by the machine-tool industry facilitated the spread of innovations, the absorption and diffusion of technology, and the continual improvement and expansion of production capabilities. Was this component of the American system successfully planted in the Russian garden?

The Stunted Sapling in the Russian Garden: The Machine-tool Industry

The genius of the left-handed craftsman in Leskov's story is that he and his Tula mates could shoe the English toy steel flea without the benefit of superior education, training, and tools. That is to say, the Russian Everyman can compensate for the absence of manmade and institutional "tools" by the possession of native genius and resourcefulness. This alleged ability of Russians to make do without the advantages possessed by the artificially endowed Europeans was celebrated not just on the pages of Russian literature and Slavophile philosophy. In one of the earliest descriptions of the Tula armory, the gunmakers were praised for their ability to make fine weapons with neither good machines nor good tools. English workers, claimed the author of this 1818 study, did not possess superior skill; they simply used superior, specialized tools. The poor Russian, continued the author, barely had a ruler, because of the "difficulty, even the impossibility, . . . to get good steel tools." Despite this disadvantage, Russian workers allegedly could make a better object than English workers in the same amount of time. Russian originality was manifested in many forms, and the anonymous author of this description of Tula recounted a story told by a government small arms inspector. Making the rounds of the workshops, it seemed to the government inspector that all the workers' hammers looked alike. It turned out that a single hammer was being passed from worker to worker in advance of the inspector. Here the author adds, "There you have a feature of the Russian people! There is nothing that these precious hands cannot do! There is nothing that these talented and gifted people cannot accomplish!"[80]

Such a casual, even scornful, attitude toward specialized tools did not signal an auspicious start for a native tool or machine-tool industry. The history of the Russian centralized machine-tool industry may be traced back to the model shop opened at Sestroretsk in 1851. The shop was established to make models of small arms and of gauges and tools, to strive for uniformity of gun parts, and to serve as a school whose graduates would spread the skills and knowledge of production processes. Under the conditions of indentured labor, it was highly unlikely that graduates would go anywhere to spread skills and production processes. Moreover, although Sestroretsk may indeed have undertaken improvements in the making of tools, gauges, and sample weapons, given Russia's centralized military procurement system, the chief beneficiary was the Artillery Department and not the other armories or the private metalworking trades. According to a history of the Sestroretsk armory, the hopes of the model shop were thwarted: the traditions of the factory remained unchanged, the factory administration resisted adoption of innovations, and the skills and production processes were not diffused throughout the Russian metal industry.[81]

This poor diffusion of skills and production processes was felt a generation later when Russia adopted breechloading rifles and revolvers. When the Tula armory was rebuilt, only a few years after the emancipation of the indentured gunmakers, there were no private machine-tool makers to take government orders for precision machinery, and the Ministry of War was forced to order abroad. Among all the skilled workers in Tula (a history of the factory estimated that in the early 1870s there were twenty thousand former gunmakers) there were few skilled mechanics and lathe operators, and even fewer tool builders. An American historian of Russian defense industries in a later period aptly characterized this as the "initial primitiveness of the Russian factories."[82]

A recent study of the St. Petersburg metalworkers at the turn of the twentieth century provides an additional clue to the troubles of the machine-tool industry. Due to the vagaries of government contracts and the weakly developed but high-priced private market, metalworking plants tended to be diversified. The typical metalworker was a general practitioner of a variety of metalworking skills. Although some specialization of skills and use of special-purpose machinery began in the arms industry in the immediate prewar years, the industry as a whole was characterized by the absence of mechanization, a reliance on hand tools, and the tendency to manufacture a diverse product line. Moreover, management was characterized by lax control, the absence of orderly work flows and routine, and the primitive use of piece rates and other incentive wage schemes.[83] Although the use of special-purpose machinery to manufacture the Berdan rifle and metallic cartridges had begun earlier than suggested by this study, if these were still the features of this all-important industry in the immediate prewar years, one can easily imagine the difficulties a generation or two earlier.

The weak machine-tool industry coupled with a dominance of the central authorities of the Artillery Department over local armory authorities inhibited research and development at the level of the plant. The fledgling machine-tool shop at Sestroretsk provided little benefit to the armory as a whole, and certainly not to other metalworking plants of the day. These features continued to plague the small arms industry when various automatic weapons systems made their appearance at the end of the century. In 1887 the Artillery Department created a special commission reminiscent of the commissions in the 1860s to consider competing designs of automatic systems, to extend invitations to specialists, and to place orders for experimental models from the government armories. More instructive than the commission created by the Artillery Department are the two commissions it did not authorize. The Officers' Sharpshooters School proposed that inventors and designers be offered the school's best facilities to organize an ad hoc research and design bureau. The superintendent of one of Tula's shops also proposed creating a special group of designers and gunmakers at the armory. "Only through close contact with the factory, and its equipment and specialists, may the inventor work productively," advised S. A. Zybin. Such early inventors who worked on designs of automatic mechanisms received little support from the Ministry of War for experimental models or facilities. Both proposals provided considerably more initiative and decision-making power to the peripheries, in this case to the armories and even to ad hoc design bureaus within the armories. The authority of the center—that is, of the Artillery Department—would have been reduced to authorizing funds and equipment. Even when designers did have permission to work on test models at the government armories, their work had to be extensively reported, permission was needed for the slightest change in design, and the experimental models were considered government property.[84]

Awareness of these deficiencies in Russian industry did not occur frequently in the contemporary foreign literature. In a 1910 study of the machine-tool trade of foreign countries, Godfrey Carden of the United States Department of Commerce and Labor called attention to the problem even though he did not offer much in the way of explanation. Carden noted that the Russian machine-tool industry was in its "infancy." Two factors, he thought, favored its development. First, the high duty on imported machine tools created favorable opportunities for domestic machine-tool plants. Second, the Russian worker was as capable as any European worker. "In several essentials," Carden wrote, "he is better than the men of some countries, for the Russian will stick to a tool and become thoroughly expert in its handling." Yet "the history of machine-tool works in Russia has, for the most part, been one of failure." Where machine-tool building had been successful, he observed, "a plant run on strictly American lines, following up-to-date American methods, can command a splendid business today in Russia." The failure lay in the manufacturing methods and organization of production tasks, and in particular in "the fact that machine shops did not specialize and did not understand the term 'shop efficiency'."[85] While it would be anachronistic to apply the standards of American efficiency experts to the Russian tool industry of the nineteenth century, Carden did call attention to the Achilles heel of Russian industry—the inattention given to specialized machine shops capable of supplying the critical research and development facilities.

If the advances Russia made in small arms production in the 1870s had been absorbed into the native metalworking industry—that is, if Russia had succeeded in domesticating the production technologies—then one would expect that during the next round of weapons modernization, system design, and adoption, Russia would have been able to rely more on native resources to design new weapons systems, retool factories for fabrication, and launch production runs more quickly. Unfortunately, a strong machine-tool industry did not grow well in the Russian garden. According to Ashurkov, the Soviet historian of the Tula armory, the greater the availability of working and measuring tools, and, of course, of skilled tool builders, meant that changes in machines, necessitated by model changes, were easier and faster in the West. "Despite the considerable advances in Russian machine building," Ashurkov rued, "its bottleneck remained the production of complex, specialized machines."[86] This weakness in Russian industry, which offset the considerable advances in the rationalization of the small arms industry, can be illustrated by a brief look at the adoption and manufacture of the next generation of small arms.

Preparing for the Next Generation of Small Arms

Not long after the single-shot Berdan rifle had become the standard infantry arm, a new generation of smaller caliber revolvers and magazine rifles with high velocity cartridges and smokeless powder was adopted all over Europe. These weapons were soon followed by the "deadly scorpion of the battlefield," the machine gun.

Russia's first magazine rifles were conversion models of the 1870 single-shot Berdan. Soviet historians give much attention to Sergei Ivanovich Mosin (1849–1902), designer of the .30 caliber magazine rifle. After graduating from the Mikhailovskii Artillery Academy in 1875, Mosin was assigned to the Tula armory, where he learned armory practice and production methods and became the chief of the tool shop. In 1882 his first designs were a mechanism to convert the Berdan into a magazine rifle of reduced caliber and a cartridge with smokeless powder. In 1889 his design for a .30 caliber, five-cartridge magazine rifle was one of many systems considered for adoption by the Artillery

Administration, and two years later the Mosin was chosen over its nearest rival, the Belgian Nagant. The Artillery Administration selected the Mosin on the grounds that it was considered simpler to operate, more durable and reliable in combat, less expensive, and more easily adapted to domestic production.[87]

When Russia had adopted the Berdan twenty years earlier, the Ministry of War made an initial weapons order abroad until the slower Russian armories could be retooled to commence domestic production. In 1891 the pattern repeated itself. Until the Russian armories could be retooled, the Ministry of War, on the eve of the Franco-Russian friendship treaty of 1891, placed an order with the French government for five hundred thousand Mosin rifles. (This order was eight times larger than the orders twenty years earlier in Hartford and Birmingham for the Berdan rifle.) At the time of the adoption of the Mosin rifle, plant capacity of Russia's three government armories was just over one quarter of a million infantry rifles per year; at the same time, plant capacity at the three French government armories was one million rifles per year, almost quadruple that of Russia. Despite the fact that French armories had just begun to retool their factories for mass production of magazine rifles, Ashurkov admitted, "Given the high level of capitalist development and the speed of technical progress, they could accomplish this quicker and easier."[88]

When Russia had adopted the Berdan rifle, its armories faced difficulties not only in the rapid reorganization of production (necessitating the "quick fix" of foreign weapons orders) but also in the provision of the appropriate precision machinery for domestic production. This pattern repeated itself a generation later. The Mosin rifle may have been simpler to make than its competitor, the Nagant, but the .30 caliber magazine rifle was considerably more complex than it predecessor, the Berdan. The small-caliber barrel and magazine required greater precision, more dies and templates, and finer detail. Its parts required tolerances of 0.001 inches and almost 1,400 discrete operations, instrument designs, and gauges. As both V. A. Tsybul'skii and L. G. Beskrovnyi admit, such precision was beyond the capabilities of Russia's machine-tool industry of the day. In addition, the lack of skilled wokers and of private small arms makers even in St. Petersburg slowed production of new dies. The shortage of native machines delayed the retooling of the Russian armories, and as a result machines as well as weapons had to be ordered abroad. Russia ordered 1,853 machines from England, France, and Switzerland; acquisition of French rifling machines was built into the French contract of 1891. (The 1,853 machines ordered abroad were twice the number ordered from Greenwood and Batley twenty years earlier to make the Berdan.)

In addition, two Russian small arms specialists were dispatched to study the production experience at the French armories.[89]

The record shows some improvement in the Russian ability to respond to new technologies at this time. In 1870 all the dies to make the Berdan had been ordered either from Colt or from Greenwood and Batley. Twenty years later, Mosin himself designed not only the rifle but also the gauges, dies, and templates for its manufacture. He was assigned to set up a tool shop and to supervise the production of the gauges and dies at the Sestroretsk armory and at the St. Petersburg cartridge factory. Because of this improvement, in the 1890s only one of seven series of dies had to be ordered abroad. Moreover, although changes in the Russian armories to convert from Berdans to Mosins involved considerable retooling, reconstruction, and expansion, the problems were worked out in less time. By 1902, almost three million Mosins had been made in Russia, and the government considered its infantry rearmament with the .30 caliber magazine to be completed.[90]

Similar patterns of foreign orders prevailed when small-caliber revolvers and machine guns were adopted. The Smith and Wesson revolver remained the standard service revolver until 1895 when the Belgian designed small-caliber Nagant was adopted. Like Smith and Wesson in the 1870s, at first the Nagant Company filled initial Russian orders; then beginning in 1898 the Tula armory began to manufacture the revolver. To help facilitate production at Tula, the Nagant Company contracted to turn over all dies, templates, and plans to Tula at the conclusion of the Nagant order. In 1896 the Russian government ordered 174 Maxim machine guns from Vickers, a British company. In 1902 it purchased a license from Vickers to make the Maxim machine gun in Russia and commenced production two years later at Tula.[91]

Several conclusions about Russian armory practice and the domestication of foreign technology can be drawn from the experience of Russian adoption and manufacture of the Mosin magazine rifle, the Nagant revolver, and the Maxim machine gun. First, as Soviet historians are quick to point out regarding the Mosin, for the first time Russian soldiers got a rifle invented and designed solely by a Russian. That, of course, could not have been said about the Mosin predecessor, the Berdan, or about the earlier conversion models such as the Carl and Krnka. Second, while the time for rearmament, or the time necessary for the absorption of a new production series, was sixteen years in the case of the Berdan, in the case of the Mosin, absorption time was reduced to ten years. Moreover, the kinks involved during the first few years of production were worked out without the full-scale restructuring of the armory that had faced the Tula armory in 1871.

Finally, that the Berdan indirectly led to its domestic successor and that a higher proportion of the gauges and dies were domestically produced suggests a certain degree of domestication of production technologies. Ashurkov concluded that the experience gained in producing the Mosin magazine rifle later aided the production of the machine gun, whose native production was mastered during the short span from 1904 to 1906.[92]

While there is evidence that the technology of the 1870s had been domesticated to a certain degree, problems still remained. The pattern of the 1860s and 1870s repeated itself: the government ordered models and small numbers of weapons from foreign arms makers for testing; after a particular system was adopted, initial orders for weapons were placed abroad while Russian armories were being retooled; precision machinery, tools, gauges, and dies were also ordered abroad; production was transferred to domestic armories. Far more Mosin rifles and Maxim machine guns were foreign made than Berdan rifles. Similarly, more machines were ordered abroad to make the Mosin than were ordered abroad to make the Berdan. Commenting on the changes in production operations in the 1890s, Tsybul'skii noted that it became necessary to substitute machine operations for hand operations. From 1882 to 1891 Sestroretsk did not acquire or make a single machine.[93] If so, this does not speak well for the degree of mechanization in the 1870s and the 1880s. Even Ashurkov, who otherwise praised Mosin's achievements, concluded that the lessons of rearmament, as in the case of the Mosin rifle, "gave no grounds for optimism regarding the capabilities of the Russian armories when compared to foreign armories."[94] Judging by the large foreign orders of weapons and machines, the difficulties in commencing domestic production, and the shortage of skilled workers, the production skills and technology acquired in the 1870s were incompletely domesticated at best. These problems get to the heart of the difficulties inherent in Russian industrial modernization—the slow absorption, diffusion, and domestication of new technologies and production processes and the poorly developed machine-tool industry.

The rearmament of the Russian army and the adoption of new technologies in small arms coincided with far-reaching changes in Russia's labor force and production processes. The transition from compulsory to free labor due to a series of imperial edicts in the 1860s proceeded with difficulty in the arms industry. Free labor performed by civilian artisans at commercially run small arms factories replaced compulsory labor performed by a caste of armorers in government armories.

Contemporary policymakers and observers regarded this transition with a mixture of enthusiasm and trepidation. Minister of War Miliutin argued that compulsory labor held back improvements in production processes. The commanding officers of the armories, essentially government employees, were allegedly bogged down in paperwork and were never involved in engineering or production problems.[95] Not being able, or not wanting, to lay off workers, armory administrators had no more incentive than did workers to mechanize. Russian defenders of a division of labor also referred to the shortage of skilled labor in the gunmaking trade. In the eyes of manufacturers, such a shortage, as had occurred in America, virtually held the manufacturers hostage to the work habits of the skilled artisans. It was widely assumed by Russians and by foreign observers that workers having the legal status of freely hired civilian artisans would work better than those having the status of serf or soldier. Indentured armorers were degraded, took no responsibility for their work, and had no incentive to innovate. In the words of a visiting English engineer, laborers kept "to the ways of their forefathers and are doing now exactly as was done five and twenty years ago."[96]

Emancipation solved one problem but created two new ones. Labor no longer bound to the armorer caste and no longer indentured to the factory could come and go. Skilled mechanics went not only from shop to shop within a community but also from one place to another, often traveling long distances. According to Chebyshev, at best wary of the efficacy of free armorers, the skilled artisans

> have been around for years and they know very well their value. . . . At the least expression of dissatisfaction on the part of the owners, they walk off their jobs and go to a different shop, knowing full well that they are in demand and will be eagerly taken wherever they go.[97]

At the same time, wage laborers not of the armorer caste made supervision, especially of the critical final fitting and assembly of the gun, more difficult. This was particularly true at Tula where such a large part of the work was performed at home or at dispersed shops.[98] Furthermore, Chebyshev argued, opening more shops to speed up the process of rifle conversion would only compound the problem. In addition, the prospect of making up for the shortfall in native skilled labor by "filling the armory with first class tool makers from other countries" was not well received.[99]

The solution to the problem of supervision seemed to be to discontinue the manual tasks performed at home, to introduce machines, and to centralize the production processes in the factory. Since the

private contractor then leasing the Tula armory, not unlike the paternalistic and parochial superintendents of Harpers Ferry, was not in-
clined to undertake such a far-reaching reorganization, the government reposssessed the armory when the lease expired. Only then was
the Tula armory ready to receive imported technology. Thus the legal
and caste supervision of the old system was replaced by the technical
supervision of the new factory system. As a result, in the government
armories, Russia made the transition from serf to machine operative
relying only briefly on independent craftsmen.

Russian pride and great power ambitions provided an imperative
for the importation and domestication of technology. The Russian government made a great effort to outfit its armories with the latest machinery so that infantry arms could be domestically produced. However, it was one thing to import a series of machines; it was quite
another to domesticate completely new technologies. That process
was far more complex and was by no means purely technical. In the
process of technological diffusion, the state undertook the rationalization of production in an era of uncertain labor-management relations.
Russian armory practice provides the paradigm for innovation and
technological change.

CONCLUSION

The State, Technology, and Labor in the Russian Small Arms Industry

● THE NINETEENTH CENTURY BEGAN THE age of the nation in arms, of firepower, and of the machine. Beginning in mid-century, governments feverishly rearmed infantries with long-range breech-loading rifles manufactured in mass quantities. The increase in fire-power brought about by the new weaponry everywhere posed a chal-lenge to military tacticians and commanders. Although conservative commanders defended the values of heroism, discipline, and will per-sonified in the cult of "cold steel," loose, extended, skirmish-line tac-tics began to replace the assault of infantry columns in closed forma-tion, and open formations placed a premium on individual initiative. The strength of armies could be measured no longer only by the number of men but increasingly by the firepower available, by the number, deployment, and capabilities of soldiers using the new weapons sys-tems, and by the productive capacities of native arms industries.

The progress associated with the nineteenth century began some-what later on the peripheries of Western Europe and North America. The age of the nation in arms, of firepower, and of the machine began in Russia, for example, after the Crimean War. But the age began with a vengeance as the nation underwent rapid changes in its social, legal, educational, and military structures. By the mid-1870s, Russia was on the road to becoming a nation in arms, a nation of firepower, and a nation of machines. Russia had adopted long-range breechloading infantry rifles and had begun to manufacture them in government ar-mories. Thus on the eve of a new round of innovations triggered by the use of high velocity magazine cartridges, smokeless powder, and the machine gun, Russia had acquired the latest technology in small arms. We can now assess the impact of the American system of manufactures in the Russian firearms industry and the roles of Russian government, technology, and labor in the modern age.

In Russia, as elsewhere, the motivations behind the government's decision to adopt machine-made breechloading rifles were largely based on factors of military supply. The increasing tactical importance of firepower, the implementation of universal military service and the reserve system, and the changes in firearms systems all increased the demand for serviceable weapons in the regiments. Since it was difficult to make guns with parts that were uniform in strength and durability, rifles were made with parts sufficiently uniform for soldiers to repair their weapons in the field. Moreover, the rapid pace of changes in firearms systems strained Russia's technical and financial ability to retool and to commence domestic manufacture of new weapons quickly.

The particular needs of the infantry and the nation's productive capacities determined the choice of new arms technologies and the pattern of adoption of new weapons systems. Because of the large, and primarily peasant, infantry and the geographical dispersion of the Russian army, the official infantry arm had to possess two essential military features—durability and convenience of maintenance. Repeatedly in the military sources these features are placed ahead of ballistic qualities such as rate of fire, range, and accuracy in appraisals of weapons systems. Because of Russia's backwardness, particularly, as Miliutin suggested, in engineering and machine-building, system adoption was slow, difficult, and costly. It seemed most rational, accordingly, to take advantage of the experience already gained in foreign countries and send military agents abroad. And where abroad? This choice was determined by design and production considerations. Once the decision had been made in 1866 to adopt the metallic cartridge, the Russians based their small arms systems on interchangeability of cartridges. It was then natural to turn to the United States, generally recognized in the post–Civil War period as the leader in the development of metallic cartridges. Once a new system was tested and adopted, the government placed initial orders in the United States and, to a lesser extent, in England. Thus, the Berdan-Gorlov metallic cartridge, the Berdan-Gorlov "Russian rifle," and the Smith and Wesson "Russian revolver" were designed and initially produced in America.

It is the nature of innovation that new technology is more appealing if it has similarities with designs or processes already in use. Most people rarely leap into the unknown. The metallic cartridge and the breech system designed by Berdan and modified by Gorlov and Gunius were similar to muzzleloader conversion systems already in use in Russia. In adopting the Berdan, the Russian government was not making a leap into the technological unknown; on the contrary, the Berdan system was appealing because it was familiar. Also, according to the English mechanics Nasmyth and Prosser, Colt machinery fit the needs of the Russian plan. When Colt machinery was first ordered during the Crimean War, the Russian government sought the technologically familiar, not the exotic.

It has been argued that the Russian government and Russian officers were highly conservative in their adoption of new weapons.[1] To be sure, the changes from muzzleloaders to breechloaders and from paper to metallic cartridges were held back by the belief that the soldier could not properly handle the new weapon and would use too much ammunition. In addition, Russian officers had attached little importance to firepower in battle, relying instead on the infantry bayonet charge, the traditional shock weapon of Europe's military elites. Russian officers feared the loss of control over soldiers trained to take the initiative and to think in battle. The Russian military and government acted conservatively, but what major military did not? Lewis Mumford aptly identifies the irony: although war is the primary stimulus to develop new technology, military men are the most reluctant to adopt it. Since military men regard themselves as the ultimate guardians of national security, they are cautious not to jeopardize that security. As Dennis Showalter points out, innovations are costly and time consuming; new weapons must demonstrate that they represent more than marginal improvements over their predecessors.[2] Certainly the Russian government treated technological innovation with extreme caution; the "firearms tragedy," as Miliutin called the protracted adoption of conversion models, demonstrated this caution.[3] But it was due less to the innate caution of military men than to the problems posed by a huge and technically untrained army and by perennial difficulties in finance and supply.

In theory, the Russian government could have adopted modern firearms without importing and domesticating the technology. Borrowing system designs abroad was neither remarkable nor costly since design technology was not confined by national borders. The great cost involved in borrowing manufacturing technology—importing machinery and equipping government armories—could have been saved, leaving funds to purchase additional weapons. Moreover, because engineering was poorly developed and skilled gunmakers were scarce in Russia, foreign orders were actually no more expensive than domestic orders and were usually filled more expeditiously.[4] Foreign purchases of weapons to supplement existing supplies during wartime were common, and even the more advanced industrial nations such as England and the United States had to find additional suppliers abroad when necessary. Turkey was virtually dependent on foreign suppliers during the Russo-Turkish War and did not fare the worse for it. If technical and economic factors had been the sole consideration in the procurement of weapons, there is no reason that Russia, too, could not have purchased weapons without importing the technology.

But foreign weapons orders were limited. The Russian government had decided that it was not content merely to purchase arms from the United States, even if such arms were manufactured at the best factories of the day. Russia, after all, was neither Egypt nor the Papal States. The Russian government was determined to be self-sufficient in defense industries and to mass produce military small arms quickly with a small number of skilled gunmakers and effective administrators. The shortcomings of the nation's production capabilities coupled with the desire to be independent in arms production dictated another essential feature of system selection: simplicity of domestic manufacture. The opportunities to circumvent tradition-bound labor and management offered by mechanized mass production dictated the adoption of the American system of manufactures.

The military and supply motives for adopting machine-made firearms overshadowed purely economic considerations such as increase in the volume of output, reduction in unit costs, and economies of scale. Indeed, mechanization and armory reorganization caused initial *dis*economies. For example, while in the long run the volume of output promised to be greater with machine-made arms, the initial setup of a series of machines meant that no arms would be produced for a period of time. At first, this seemed to surprise the Russians, who may have expected miracles from machine production.[5] Likewise, the long-term reductions in unit costs and economies of scale of employing machinery were no more apparent to Russian armory directors than they had been to wary American arms makers a half century earlier. In the primacy of military over economic considerations, the experience of the Russian government was very much like that of other governments desiring to be self-sufficient in arms production.

Judged, then, in terms of military supply, Russia's rearmament and its progress in developing its native small arms industry were a considerable achievement. By the early 1870s, Russian small arms production finally had been converted to the centralized armory system perfected a generation or two earlier by government armories and private arms manufacturers in New England. Production had been centralized in a single factory under strict supervision and control; subcontracting in the private homes and workshops of individual gunmakers had been nearly eliminated; many production processes had been mechanized using the sequential operation of special-purpose machines; the increasing number of operations in the production of gun components had subdivided shops and diluted skills; jigs and fixtures had been adopted that permitted a high degree of interchangeability, with only a small amount of filing and finishing to be done by hand; and tool shops on the factory grounds had begun to specialize in that most critical aspect of the metalworking industry—the production of machines to make other machines. If the advances Russia made in small arms production in the preceding decade had been absorbed in the native metalworking industry—that is, if Russia had succeeded in domesticating the production technologies—then one would expect that during the next wave of weapons modernization, system design, and adoption Russia would have been able to rely more on native resources to design new weapons systems, to retool factories for the fabrication of weapons, and to launch production runs more quickly.

However, judging by the large foreign orders of weapons and machines in the 1890s, as well as by accounts of the difficulties in commencing domestic production, the lack of managerial control, the continued shortage of skilled workers, and the low degree of specialization of product line, of mechanization, and of labor in metalworking, the production skills and technology acquired in the 1860s and 1870s were incompletely domesticated. Since the production processes at Tula earlier in the century had likewise not been fully diffused, the pattern of technological importation and adoption in the 1890s was remarkably similar to that of the previous generation. As a British artillery officer observed in 1867, "At the present time when a new arm must be introduced to take the place of the old smooth bore, deficiencies in the military establishments of Russia are most apparent."[6] These deficiencies get to the heart of difficulties inherent in Russian industrial modernization: the absorption, diffusion, and domestication of new technologies and production processes, and the relationship between government and labor.

Until the middle of the nineteenth century, Russian society was a collection of self-sufficient enclaves. The productive efforts of the manor, the peasant village, the infantry regiment, and the possessional factories were neither purchased nor organized by the market. During periods of slow technological change, the self-sufficient enclaves mobilized human and material resources well enough to support an empire and Russia's great power ambitions. The self-sufficient enclaves were well suited to the government's suspicion of the market and its desire to farm out the supervision of its subjects without committing the financial resources to long-term development. Like other self-sufficient enclaves, the government armories perpetuated labor parochialism and overmanning, inhibited the workers' sense of independence and craft pride, minimized contact with the outside world, perpetuated the imbalance between the central and local authorities, reinforced sectoral segregation, and thwarted innovation and the diffusion of new methods. Worse, the armories remained self-sufficient enclaves, not integrated into the modern and dynamic society Russia needed to preserve its great power status and native institutions. The strengths and weaknesses of the armories provide a paradigm for the dilemma of development in an age of rapid change.

The shop organization of jobs, the subcontracting system, and the large portion of work done outside the armory, particularly at Tula, fostered a decentralized organization of production. Contemporaries were divided over whether the system was good or bad for the Russian firearms industry. Although for two centuries in the West centralized work has been associated (usually by proponents) with greater productivity and innovation, now, when the world is paying increased attention to workplace decentralization, "flex-time," and other measures to alter assembly-line work, it should not be assumed that the workshop system was unproductive or not conducive to innovation.

Be that as it may, the decentralized organization of production perpetuated Russia's labor parochialism. Thus the more centralized, more modern armories trained at high cost an unskilled and unmotivated labor force; the armory that trained at low cost a more motivated labor force was the most decentralized. Even though Sestroretsk, and to a lesser extent Izhevsk, centralized work in one location, the factory system with its attendant advantages in supervision was merely grafted onto the preexisting shop organization of work. Since neither the work nor the armorers could be supervised, it was difficult for the chief engineer to enforce quality control of the work performed in numerous workshops. Although the foreman had authority over the shop, it is not clear whether he was always able to exercise that authority effectively.[7] Perhaps most disadvantageous in the long run was that gunmakers and other metal craftsmen dispersed in self-sufficient enclaves were less able to train each other and to learn from each other. This acted as a further brake on the diffusion of technology within the industry.

A preindustrial culture of craft, community, and self-sufficiency made the Russian worker, like workers elsewhere, suspicious of change. As recent studies of Russian labor have shown, skilled workers, particularly those in the metal trade, worked in small scale, unmechanized, self-sufficient shops producing a diversity of articles. A lax and undisciplined shop structure controlled the intensity of labor, reinforced the sense of a "just job," and perpetuated traditional customs and observances. For skilled workers, the limited division of labor, the importance of seniority, and the weakness of managerial influence resulted in considerable autonomy, control over the labor process, and collectivist consciousness.[8]

Craft consciousness did not permeate well into the gun trade. Under serfdom, Russia experienced a shortage of skilled artisans. Indentured to the armories, skilled gunmakers and other laborers had no incentive to become independent artisans or to develop craft pride. Many nineteenth-century studies of the Russian gun trade suggest that by and large meticulous workmanship and craft pride were wanting among gunmakers. Russian gunmakers made weapons for the army, not for commercial sale, and the Russian gun trade demanded few custom-made guns.[9] The best made gun was rarely regarded as a work of art, as it occasionally was at Liège and Birmingham. The small private market in Russia for firearms provided weak stimulus to the development of a large number of skilled gunmakers and to the development of craftsmanship. The gunmakers allegedly had no self-interest in their work, had no motivation to improve, and lacked craft pride. In addition, they were indifferent and distrustful toward innovation. Such allegations were not directed only at workers. The directors, superintendents, and other armory personnel generally had little technical training and specialized knowledge. In the words of one industry survey in 1862, "the lack of knowledge [of production] . . . is the chief scourge of all our businesses."[10] The superintendents managed the armories more as instruments of the tsar's will than as entrepreneurs seeking individual or corporate gain. Russian armory managers shared a culture of administration, not of production.

The institutional expression of the gunmakers' collectivist consciousness, the *artel'*, reflected the preference for self-sufficiency induced by serfdom and the inclination of laborers to band together to equalize sacrifice and to share adversity. As Michael Kaser argues, the *artel'* reflected an acceptance of adversity, mutual reliance, and an ethos of egalitarian collectivism rather than a culture of production and an ethos of corporate gain.[11] Although the familiar concept of craft consciousness can be applied to Russia's more independent skilled craftsmen, it is less applicable to Russia's armorers. Given the long tradition of indenture of much of Russian labor, the armorers, while isolated and small in number, in many ways may be more typical of the culture of Russian labor as a whole. *Artel'* consciousness rather than craft consciousness characterized the self-sufficient enclave of the government armories.

A growing private market and an interdependence among industries and between civilian and military sectors fostered an interindustry flow and diffusion of technology in Western Europe and America. In Russia, however, the self-sufficient enclaves minimized contact with the outside world, reinforced sectoral isolations, and perpetuated the imbalance between the central and local authorities. The geographical and sectoral isolation of the components of the Russian economy inhibited information flow, rapid reproduction and assimilation of new techniques, and spin-off technologies. In particular, a segregation of the civilian and military sectors of the economy meant that the latter made little lasting contribution to the former.

The domestic private market in small arms was virtually nonexistent, and there was never a large number of gunmakers in the private sector whose numbers could complement the government armorers. This situation was exacerbated because there were too few private factories "capable of giving assistance in time of emergency." It was commonly argued that, despite the lure of profits, private owners would have no incentive to incur the expense and risk of plant modernization after they had concluded a contract for deliveries to the state.[12] This deficiency in private entrepreneurship threw a double responsibility on the government and forced it to turn to state industries. Yet the government had given no support to private industry. Unlike the American government, which offered advances to private manufacturers, the Russian government was reluctant to part with funds, even for experimental models and facilities, to train skilled weapons designers and to organize work systematically. The need for government monopoly in the production of small arms was based on a fundamental mistrust of private manufacturers and a fear of dependence on private producers. Admittedly, the small arms industry is a special case, but the root of this fear—the mistrust of private gain and of the alleged rapaciousness of the individual entrepreneur in an economy of scarcities—went deep in Russian culture and was shared by government and society alike.

The self-sufficient enclaves, combined with a government monopoly in the small arms industry, reinforced the imbalance between the center and the periphery. The many government commissions for weapons design and testing channeled innovations to the central authorities, the Artillery Department, and the Ministry of War. Attempts to fund and support research and design work at the peripheries—that is, at the armories and related factories themselves —were consistently rejected. The government was suspicious not only of private entrepreneurial initiative but also of local initiative even at its own small arms factories. Such an imbalance between the center and the periphery stunted the growth of the machine-tool industry, of research and design facilities, and of the cross-fertilization between science and application, an important feature of modern industry. However, the Russian government did not capitalize on this imbalance between the center and the periphery. Although the government owned the armories and controlled a centralized procurement system, it showed only a belated interest in modernizing armory practice, particularly at Tula, the largest armory. Here, Russia differed from the United States, where the Ordnance Department in Washington showed unceasing interest in rationalizing production techniques at Harpers Ferry.

Although the armories had an outside consumer of their product, this consumer was the government itself. To be sure, toward the end of the nineteenth century there was an increasing interdependence between military and civilian industries, especially in munitions, artillery pieces, and heavy industry in St. Petersburg; and a private market for small arms did appear. However, the military and civilian industries were segregated in the important decades of the 1860s and the 1870s. The self-sufficient enclaves did not encourage outsiders who could bring innovations. Examples of influential and innovative outsiders in the government armories are rare. This was particularly detrimental to the process of technical innovation in metalworking and small arms. Russia, however, had a stunted machine-tool industry and, accordingly, did not experience the "technological convergence" that characterized the development of precision machine tools and sophisticated production processes in the United States during the first half of the nineteenth century. One reads the histories of technological innovation and of individual armories in vain for accounts of cross-fertilization between arms makers and tool builders or between the caste of gunmakers and metalworkers in civilian establishments. Thus a variety of institutional barriers impeded the diffusion and assimilation of new technologies.

Given the many dilemmas of development, it is remarkable that Russia kept pace with the rapid changes in weapons and weapons production. Even though the self-sufficient enclaves thwarted technological diffusion, the government monopoly in small arms production nevertheless offered some important advantages in technological adoption. Government resources could be mobilized quickly and directed toward a single task, and on occasion a competent and resourceful official, such as Colonel Gorlov, could bulldoze his way through institutional barriers. If resources were mobilized in the right direction, a "quick fix" could be achieved. We have already seen that the Ministry of War had decided to renationalize the Tula armory. The

Notbek Commission had correctly seen the need for both precision machines and centralized armory practice of, in its words, "a single production enterprise." To reproduce the organization of work at the Colt factory in Russia and to "establish in Russia a mechanized armory on the American model"[13] provided an opportunity to revolutionize Russian small arms production and to effect a transition from serfs to machine operatives without an intervening independent craft stage. As the Colt facilities allowed American independent subcontractors to improve system design and production processes, so too they provided the Russians an important learning laboratory at a time when Russian armory practice was being reexamined. The emphasis in New England armory practice on process over product provided the learning model that had previously eluded the Russians in their efforts to innovate in small arms production. Consequently, by government decree the Tula armory became a centralized, mechanized factory producing a large series of rifles using sequential, special-purpose machines and tools. As machine operatives allowed American manufacturers and engineers to compensate for the shortage of skilled workers, so too in Imperial Russia machine operatives allowed the government to compensate for the self-sufficient enclave of skilled gunmakers. Once the American system of manufactures had been planted in the Russian garden, the Russian government returned to traditional, and less costly, suppliers in Europe for subsequent borrowing of design and machines. The resulting relations of production constituted a form of state capitalism in the small arms industry, and, ironically, the Russians found their model of state capitalism in American armory practice.

Epilogue

The Soviet Case and the Russian Pattern

At first glance it does not seem likely that the Russian small arms industry would have had much impact on the development of Soviet industry. This was a small and rather isolated sector of Russian industry. Besides, the peculiar needs of the Soviet state and the Stalinist development model would be the logical source for the features of Soviet industry. However, that the implementation of the Stalinist development model of centrally planned production and procurement created a militarized economy and society suggests that the source of many features of Soviet economic practice may be found not in Soviet ideology but in the legacy of prerevolutionary military industries.

The industrial development model frequently chosen for individual enterprises during the First Five-Year Plan had its precedent in prerevolutionary armory practice. In his study of Soviet metal fabricating techniques, David Granick identifies two industrial organization models available to Soviet planners drawing up the First Five-Year Plan in the late 1920s. The first model, that of the West European industrial tradition, based industrial processes on a high degree of craftsmanship and on an acceptance of a comparative lack of specialization and economies of scale. The second model represented the American industrial tradition, dating back to New England armory practice. The important features of this model, according to Granick, were mass production of large lot sizes, standardization of product, continuous flow technology, extensive use of jigs and fixtures, and a low input of manual craft skills.[14] Although many Soviet planners viewed the West European model as more appropriate for industry as it existed in 1929, and although integration of the American model presented formidable difficulties, the American model offered a solution to certain problems of development. The larger the lot size, the less equipment changeover required—an important consideration given the lack of capital. The continuous flow of work from one machine to another, requiring tight schedules, rigid quality standards, and additional supervisory personnel, promised an organizational method well suited to a semiskilled workforce. A high degree of subdivision of operations shortened the training of this semiskilled workforce, clearly an important consideration given the pace of industrialization contemplated.

> In the metal-fabricating industries, the American model was incorporated not so much in terms of particular types of specialized equipment . . . as in terms of production organization. Mass production and large lot sizes, with the necessary concomitant of plant specialization, on the one hand, and continuous flow organization of production on the other were forms of organization which represented modernity. Although they would not necessarily provide superior results in the short run, these techniques represented the wave of the future.[15]

That the Russians regarded such practices as a peculiarly American technique of manufacturing suggests that initial elements of this technique, borrowed one half century earlier, had not been successfully diffused and had even been altogether forgotten.

Were the imported American techniques of the 1930s diffused? Because imported technology must continually be updated, it is never a substitute for domestic innovation. At the Americanized auto plant at Gorky, follow-up technology provided by Ford was not diffused, and Soviet sources indicate that new technologies at the showcase Togliatti auto plant have not been diffused or updated. A pattern of returning to western suppliers for new vintages of technologies acquired earlier suggests a failure to develop native technology.[16] According to a study of Soviet-Western trade, "A weakness of Soviet copying is that research and development efforts often stopped with the duplication of western prototypes. Little Soviet innovation and product improvement followed and . . . Soviet efforts at improving on western designs were often unsuccessful." This feature of Soviet development has its antecedents in the Russian practice, noted by Michael Kaser, of achieving economies of scale by copying and multiplying units rather than by progressing to new levels of technology. The Russian and Soviet patterns have a common origin: the desire to avoid the risk associated with innovation.[17]

Among the many factors impeding Soviet innovation and diffusion surveyed in the current literature, three are particularly important in the context of this study. First, the lack of material incentives, the obstacles to individual initiative, and the absence of competition provide little motivation to introduce new products or processes.[18] Second, the high cost of new technology and sophisticated equipment has dictated a distinctive development pattern. On the one hand, in high priority fields, established technologies have been improved incrementally; on the other hand, existing productive capacity in the Soviet Union is withdrawn from service more slowly than in the West. As a result, goods continue to be produced even after production of new goods has begun. According to Philip Hanson, "This is not always a self-evidently inefficient thing to do, but it must often be so and it certainly is particularly characteristic of Soviet production."[19] Third, Soviet and Western studies have pointed to the undesirable effect of "departmentalism," lack of sectoral coordination, and poor information flow on the diffusion of new processes and on the perpetuation of the autarkic tendencies of enterprises.[20]

These obstacles to innovation and diffusion can be observed in the Soviet machine-tool industry and in Soviet weapons design. Automated machine-tool construction has proceeded more slowly than in the West. The poor cooperation between the machine-tool and electronics industries has held back the development of high-performance tools and perpetuated the bias for simpler, general-purpose tools with long service lives.[21] Soviet weapons designers, not unlike their tsarist predecessors, modify existing systems and introduce incremental improvements rather than design new weapons from scratch. Thus new products are familiar. Economy in the design and use of components makes development, production, and maintenance cheaper. Simplicity, rugged design, durability, and reliability, top priority for rifles more than one hundred years ago, remain the top priority for Soviet tanks today.[22]

Recent studies of Soviet technology, research and development, and technology transfer suggest that Soviet acquisitions of western technology have saved millions of dollars in research and development costs and lead time, have reduced engineering risks, and have achieved greater product performance. Compared to other borrowers of military technology, the Soviet Union has adopted a more self-reliant policy in order to reduce dependence on foreign technology.[23] However, while lag may be least apparent in the research and experimental stages, the absorption gap remains greatest at the diffusion stage. Of course, reform is on the way in the form of the new enterprise law and self-financing designed to give plant managers more autonomy and incentive to increase productivity. It is unlikely that this change will be any more of a panacea than was the analogous leasehold management system in the imperial government armories. The chief barrier to effective assimilation of new technology remains the economic system of the Soviet Union, particularly its inflexibility and self-sufficiency, the same obstacles to innovation and technological diffusion more than a century ago.

Abbreviations

A. Zh.	*Artilleriiskii zhurnal*
CHS	Connecticut Historical Society
CPFAM	Colt Patent Fire Arms Manufacturing Company
CSL	Connecticut State Library
M.S.	*Morskoi sbornik*
NA	National Archives
O.S.	*Oruzheinyi sbornik*
PRO	Public Record Office
RG	Record Group
R.I.	*Russkii invalid*
T.C.	*Technology and Culture*
Uchenye zapiski T.G.P.I.	*Uchenye zapiski Tul'skogo gosudarstvennogo pedagogicheskogo instituta*
V.L.U.	*Vestnik Leningradskogo Universiteta*
V.S.	*Voennyi sbornik*
Zapiski R.T.O.	*Zapiski Russkogo tekhnicheskogo obshchestva*

Notes

Chapter 1. Introduction

1. This has been most recently argued in Teodor Shanin, *Russia as a "Developing Society." The Roots of Otherness: Russia's Turn of the Century*, 2 vols. (New Haven and London, 1986), 1: xi.
2. J. F. C. Fuller, *Armament and History* (New York, 1945), 117.
3. Alfred J. Rieber, *The Politics of Autocracy: Letters of Alexander II to Prince A. I. Bariatinskii, 1857–1864* (Paris, 1966), 24–30.
4. "Vzgliad na sostoianie russkikh voisk," *Voennyi sbornik*, 1 (1858): 1.
5. Ibid. See also V. G. Fedorov, *Vooruzhenie russkoi armii v Krymskuiu kampaniiu* (St. Petersburg, 1904), 5.
6. Bruce Menning, *Bayonets Before Bullets: The Organization and Tactics of the Imperial Russian Army, 1861–1905* (Fort Leavenworth, Kans., 1984), ix–x.
7. Walter Pintner, "The Burden of Defense in Imperial Russia," *Russian Review* 43, 3 (July 1984): 232–34, 241.
8. Quoted in Fedorov, *Vooruzhenie russkoi armii*, 135–36.
9. Ibid., 155–56.
10. Rieber, *The Politics of Autocracy*, 96. On the influence of long-term economic and technological changes, see the stimulating paper by Peter Gatrell, "The Meaning of the Great Reforms in Russian Economic History," prepared for the conference "The Great Reforms in Russian History, 1861–1874," University of Pennsylvania, 25–28 May 1989.
11. S. B. Saul, "The Nature and Diffusion of Technology," in A. J. Youngson, ed., *Economic Development in the Long Run* (London, 1972), 53–54.
12. Nathan Rosenberg, "Technological Interdependence in the American Economy," *Technology and Culture* 20, 1 (January 1979): 41, 46; J. E. S. Parker, *The Economics of Innovation: The National and Multinational Enterprise in Technological Change* (London, 1974), 102; Lawrence A. Brown, *Innovation Diffusion: A New Perspective* (London, 1981), 6–7; Thane Gustafson, *Selling the Russians the Rope? Soviet Technology Policy and U.S. Export Controls*, Rand Publications R-2649-ARPA (Santa Monica, Calif., 1981), 74.
13. Merritt Roe Smith provides a useful survey of the issues and the secondary sources on the interface between the American military and industry in

his introduction to Merritt Roe Smith, ed., *Military Enterprise and Technological Change: Perspectives on the American Experience* (Cambridge, Mass., 1985). I owe an intellectual debt to the following works: Merritt Roe Smith, *The Harpers Ferry Armory and the New Technology: The Challenge of Change* (Ithaca, N.Y., 1977); Nathan Rosenberg, *Technology and American Economic Growth* (New York, 1972); Nathan Rosenberg, *The American System of Manufactures* (Edinburgh, 1969); Paul J. Uselding, "An Early Chapter in the Evolution of American Industrial Management," in Louis P. Cain and Paul J. Uselding, eds., *Business Enterprise and Economic Change* (Kent, Ohio, 1973), 51–84; Paul J. Uselding, "Studies of Technology in Economic Hisory," in Robert E. Gallman, ed., *Recent Developments in the Study of Business Economic History: Essays in Memory of Herman E. Krooss* (Greenwich, Ct., 1977); Paul J. Uselding, "Elisha K. Root and the American System," *T. C.* 15, 4 (October 1974): 543–68; Robert A. Howard, "Interchangeable Parts Reexamined: The Private Sector of the American Arms Industry on the Eve of the Civil War," *T. C.* 19, 4 (October 1979): 633–49; and David A. Hounshell, *From the American System to Mass Production, 1800–1932: The Development of Manufacturing Technology in the United States* (Baltimore, 1984).

14. For a discussion of the various definitions of this term, see Hounshell, *From the American System,* 15–25, 35, 46–50. See also Felicia Deyrup, *Arms Makers of the Connecticut Valley* (Northampton, Mass., 1948), 3–5; Uselding, "Studies of Technology," 168.

15. Smith, *The Harpers Ferry Armory,* 329.

16. Ibid., 283, 324–25; Rosenberg, *Technology and American Economic Growth,* 91, 95.

17. H. Van der Hass, *The Enterprise in Transition: An Analysis of European and American Practice* (London, 1967), 5, 18, 167. See also Cyril Falls, *A Hundred Years of Warfare* (New York, 1953), 18.

18. Mira Wilkins, *The Emergence of the Multi-national Corporation: American Business Abroad from the Colonial Era to 1914* (Cambridge, Mass., 1970), 29, 37, 66, 75–76. See also James Everett Katz, ed., *Arms Production in Developing Countries: An Analysis of Decision Making* (Lexington, Mass. and Toronto, 1984), 5–9.

19. Wilkins, *Emergence,* 102–3. On the control of foreign business activity, see Walther Kirchner, "Russian Entrepreneurship and the Russification of Foreign Enterprise," *Zeitschrift für Unternehmensgeschichte* 26 (1981): 89; and Frederick V. Carstensen, *American Enterprise in Foreign Markets: Studies of Singer and International Harvester in Imperial Russia* (Chapel Hill, N.C., 1984), 7. For more on entrepreneurship in Russia, see Thomas C. Owen, *Capitalism and Politics: A Social History of the Moscow Merchants* (Cambridge, 1981); Alfred J. Rieber, *Merchants and Entrepreneurs in Imperial Russia* (Chapel Hill, N.C., 1981); and Michael Kaser, "Russian Entrepreneurship," in *The Cambridge Economic History of Europe* vol. 7, pt. 2 (Cambridge, 1978), 416–93.

20. John S. Curtiss, *The Army of Nicholas I* (Durham, N.C., 1965); G. P.

Meshcheriakov, *Russkaia voennaia mysl' v XIX v.* (Moscow, 1973); L. G. Beskrovnyi, *Russkaia armiia i flot v XIX v.* (Moscow, 1973); A. A. Strokov, *Istoriia voennogo iskusstva*, 2 vols. (Moscow, 1965); V. A. Avdeev and P.A. Zhilin, *Russkaia voennaia mysl', konets XIX-nachalo XX v.* (Moscow, 1982); A. V. Fedorov, *Russkaia armiia v 50–70 kh gg. XIX v.* (Leningrad, 1959).

21. P. A. Zaionchkovskii, *Voennye reformy*; idem, "Voennye reformy D. A. Miliutina," *Voprosy istorii* 2 (1945): 3–27; Forrestt Miller, *Dmitrii Miliutin and the Reform Era in Russia* (Nashville, Tenn., 1968). For example, Miller's study of Miliutin completely neglects the important area of arms, tactics, and the defense industries.

22. Allan K. Wildman, *The End of the Imperial Russian Army*, 2 vols. (Princeton, 1980, 1987); Dietrich Beyrau, *Militär und Gesellschaft in vorrevolutionaren Russland* (Colgne and Vienna, 1984); William C. Fuller, *Civil-Military Conflict in Imperial Russia, 1881–1914* (Princeton, 1985); John Bushnell, *Mutiny Amid Repression: Russian Soldiers in the Revolution of 1905–1906* (Bloomington, Ind., 1985); Menning, *Bayonets Before Bullets*; Peter Von Wahlde, "Military Thought in Imperial Russia," Ph.D. diss. (Indiana University, 1966); E. Willis Brooks, "Reform in the Russia Army," *Slavic Review* 43 (Spring 1984): 63–82. For partial exceptions, see Jacob Kipp, "The Russian Naval Ministry and the Introduction of the Ironclad: An Aspect of Russian Economic Development, 1862–1867," unpublished paper; and Edward Ralph Goldstein, "Military Aspects of Russian Industrialization: The Defense Industries, 1890–1917," Ph.D. diss. (Case Western Reserve University, 1971).

23. Geroid T. Robinson, *Rural Russia Under the Old Regime* (Berkeley, 1967); Terence Emmons, *The Russian Landed Gentry and the Peasant Emancipation of 1861* (Cambridge, 1968); Daniel Field, *The End of Serfdom: Nobility and Bureaucracy in Russia, 1855–1861* (Cambridge, Mass., 1976); Rieber, *The Politics of Autocracy*; P. A. Zaionchkovskii, *Otmena krepostnogo prava v Rossi*, 3d. ed. (Moscow, 1968), and *Provedenie v zhizni krest'ianskoi reformy 1861 g.* (Moscow, 1958).

24. Theodore Von Laue, *Sergei Witte and the Industrialization of Russia* (New York, 1963); Rieber, *Merchants and Entrepreneurs*; Owen, *Capitalism and Politics*; Peter Gatrell, *The Tsarist Economy, 1850–1917* (New York, 1986); Dietrich Geyer, *Russian Imperialism* (Leamington Spa, 1986); Peter Gatrell, "Industrial Expansion in Tsarist Russia, 1908–1914," *Economic History Review* 35, 1 (1982): 99–110.

25. John P. McKay, *Pioneers for Profit: Foreign Entrepreneurship and Russian Industrialization, 1885–1913* (Chicago, 1970); Walther Kirchner, "The Industrialization of Russia and the Siemens Firm, 1853–1890," *Jahrbücher für Geschichte Osteuropas* 22, 3 (1974): 321–27; idem, "Western Businessmen in Russia: Practices and Problems," *Business History Review* 38 (1964): 315–27; idem, *Studies in Russian-American Commerce, 1820–1860* (Leiden, 1975); idem, "Russian Entrepreneurship and the Russification of Foreign Enterprise," *Zeitschrift für Unternehmensgeschichte* 26 (1981): 79–103; idem, "One Hundred Years of Krupp and Russia, 1818–1918," in

Vierteljahrschrift für Sozial- und Wirtschaftsgeschichte 69 (1982): 75–108; idem, *Die Deutsche Industrie und die Industrialisierung Russlands, 1815–1914* (St. Katharinen, 1986); Olga Crisp, *Studies in the Russian Economy before 1914* (London, 1976); Carstensen, *American Enterprise*; E. R. Goldstein, "Vickers Ltd. and the Tsarist Regime," *Slavonic and East European Review* 58 (1980): 561–71; and G. Jones and C. Trebilcock, "Russian Industry and British Business, 1910–1930: Oil and Armaments," *Journal of Economic History* 11 (1982): 61–103.

26. V. V. Mavrodin and P. Sh. Sot, "Sovetskaia istoriografiia otechestvennogo oruzhiia XIX-nachale XX v.," *V. L. U.* 14 (1976): 45.

27. Mikhail Tugan-Baranovskii, *Russkaia fabrika v proshlom i nastoiashchem: Istoricheskoe razvitie russkoi fabriki v XIX v.,* 3d ed. (Moscow, 1907); P. A. Khromov, *Ekonomicheskoe razvitie Rossii: Ocherki ekonomiki Rossii s drevneishikh vremen do Velikoi Oktiabr'skoi revoliutsii* (Moscow, 1967); A. G. Rashin, *Formirovanie rabochego klassa Rossii: Istoriko-ekonomicheskie ocherki* (Moscow, 1958); P. I. Liashchenko, *History of the National Economy of Russia to the 1917 Revolution,* trans. L. M. Herman (New York, 1949); M. S. Volin and Iu. I. Kir'ianov, eds., *Istoriia rabochego klassa SSSR,* vol. 1 (Moscow, 1983).

28. F. Sartisson, *Beiträge zur Geschichte und Statistik des russischen Bergbau-und Huttenwesens* (Heidelberg, 1900); M. Kashkarov, *Statisticheskii ocherk khoziaistva i imushchestvennogo polozheniia krest'ian orlovskoi i tul'skoi gubernii* (St. Petersburg, 1902); George Cleinow, *Beiträge zur lage der hausindustrie in Tula* (Leipzig, 1904); G. Bakulev and D. Solomentsev, *Promyshlennost' Tul'skogo ekonomicheskogo raiona* (Tula, 1960). For partial exceptions, see Ivan F. Afremov, *Istoricheskoe obozrenie tul'skoi gubernii* (Moscow, 1850); and A. S. Britkin, *The Craftsmen of Tula: Pioneer Builders of Water-Driven Machinery* (Jerusalem, 1967). Two widely cited contemporary studies of Russian industry ignored the arms industry: Ludwik Tegoborski, *Commentaries on the Productive Forces of Russia,* 2 vols. (London, 1855); and Gerhart von Schulze-Gävernitz, *Volkswirtschaftliche studien aus Russland* (Leipzig, 1899). Even studies of the iron, metallurgical, and machine industry neglect the arms industry: V. K. Iatsunskii, "Krupnaia promyshlennost' Rossii v 1790–1860 gg.," in M. K. Rozhkova, ed., *Ocherki ekonomicheskoi istorii Rossii pervoi poloviny XIX v.* (Moscow, 1959); S. G. Strumilin, *Istoriia chernoi metallurgii v SSSR* (Moscow, 1954); idem, *Chernaia metallurgiia v Rossii i v SSSR: Tekhnicheskii progress za 300 let* (Moscow, 1935); Ia. S. Rozenfel'd and K. I. Klimenko, *Istoriia mashinostroeniia SSSR s pervoi poloviny XIX v. do nashikh dnei* (Moscow, 1961); M. A. Pavlov, ed., *Metallurgicheskie zavody na territorii SSSR s XVII do 1917 g.* (Moscow, 1937). In an otherwise stimulating discussion about "decoupling" serfdom from technological backwardness, Thomas Esper neglects the arms industry. See "Industrial Serfdom and Metallurgical Technology in 19th-Century Russia," *T. C.* 23 (October 1982): 583–608.

29. Olga Crisp, "Labor and Industrialization in Russia," *Cambridge Economic History of Europe,* vol. 7, pt. 2 (Cambridge, 1978): 308–415; Victoria Bonnell,

Roots of Rebellion: Workers' Politics and Organizations in St. Petersburg and Moscow, 1900–1914 (Berkeley, 1983); Victoria Bonnell, ed., *The Russian Worker* (Berkeley, 1984); Reginald E. Zelnik, trans. and ed., *A Radical Worker in Tsarist Russia: The Autobiography of Semen Ivanovich Kanatchikov* (Stanford, 1986); Heather Hogan, "Labor and Management in Conflict: The St. Petersburg Metalworking Industry, 1900–1914," Ph.D. diss. (Michigan, 1981); and idem, "Industrial Rationalization and the Roots of Labor Militance in the St. Petersburg Metalworking Industry, 1901–1914," *Russian Review* 42, 2 (April 1983): 163–90.

30. V. N. Ashurkov, *Kuznitsa oruzhiia: Ocherki po istorii Tul'skogo oruzheinogo zavoda* (Tula, 1947); idem, "Russkie oruzheinye zavody v 40–50kh gg. XIX v.," *Voprosy voennoi istorii Rossii* (Moscow, 1969); A. A. Aleksandrov, *Izhevskii zavod: Nauchno-populiarnyi ocherk istorii zavoda, 1760–1917* (Izhevsk, 1957); V. V. Mavrodin, "K voprosu o perevooruzhenii russkoi armii v seredine XIX v.," in *Problemy istorii feodal'noi Rossii* (Leningrad, 1971); idem, ed., *Rabochie oruzheinoi promyshlennosti v Rossii i russkie oruzheiniki v XIX–nachale XX v.* (Leningrad, 1976); V. V. Mavrodin and Val. V. Mavrodin, *Iz istorii otechestvennogo oruzhiia: Russkaia vintovka* (Leningrad, 1981).

31. P. A. Zaionchkovskii, *Voennye reformy, 1860–1870* (Moscow, 1952), 137. See also G. P. Meshcheriakov, *Russkaia voennaia mysl'*, 171; and N. I. Gnatkovskii and P. A. Shorin, *Istoriia razvitiia otechestvennogo strelkogo oruzhiia* (Moscow, 1959), 79.

Chapter 2. Firearms in the Industrial Age

1. Maurice Matloff, "The Nature and Scope of Military History," in Russell F. Weigley, ed., *New Dimensions in Military History* (San Rafael, Calif.,1975), 390.

2. J. F. C. Fuller, *Armament and History*, (New York, 1945) 110–13; Rosenberg, *Technology and American Economic Growth* (New York, 1972), 168; Leonid Tarassuk and Claude Blair, eds., *The Complete Encylopedia of Arms and Weapons* (New York, 1979), 221.

3. United States Senate, 36th Congress, 1st session, Executive Document no. 60, *Military Commission to Europe in 1855 and 1856: Report of Major Alfred Mordecai of the Ordnance Department* (Washington, D.C., 1860), 172 (hereafter cited as Mordecai, *Report*); B.H. Liddell Hart, "Armed Forces and the Art of War: Armies," in *The New Cambridge Modern History*, 12 vols. (Cambridge, 1957–1962), 10: 305; "Rifle," *Encyclopedia Britannica*, 11th ed., 23: 326–27.

4. Fuller, *Armament and History*, 110–11; Joseph Schoen, *Das gezogene Infantrie-Gewehr. Kurze Darstellung der Waffensysteme der Neuzeit u. ihrer Anwendung in der Armeen Europas* (Dresden, 1855), translated into Russian as *Pekhotnoe nareznoe ruzh'e. Kratkoe obozrenie noveishikh sistem oruzhii i primeniia ikh k vooruzhenii pekhoty v razlichnykh evropeiskikh gosudarstvakh* (St. Petersburg, 1858), 209–10; Mordecai, *Re-*

port, 172. Mordecai appended to his report an English translation of Schoen's work, entitled *Rifled Infantry Arms*, from which the following citations will be drawn.

5. Schoen, *Rifled Infantry Arms*, 195; J. R. Newman, *The Tools of War* (New York, 1942), 44. The needle gun is given considerable attention in Dennis Showalter, *Railroads and Rifles* (Hamden, 1975).

6. Mordecai, *Report*, 172; Tarassuk and Blair, *Complete Encyclopedia of Arms and Weapons*, 224–25; Schoen, *Rifled Infantry Arms*, 195; Showalter, *Railroads and Rifles*, 237–38.

7. Hart, "Armed Forces," 305; W. H. B. Smith, *Small Arms of the World*, 10th ed. (Harrisburg, Pa., 1973), 29; Tarassuk and Blair, *Complete Encyclopedia of Arms and Weapons*, 224–25.

8. Fuller, *Armament and History*, 112–13; Smith, *Small Arms of the World*, 43–46; Tarassuk and Blair, *Complete Encyclopedia of Arms and Weapons*, 114–15; "American Industries No. 75: The Firearms Manufacture," *Scientific American* (3, September 1881): 148.

9. Smith, *Small Arms of the World*, 43–46; Tarassuk and Blair, *Complete Encyclopedia of Arms and Weapons*, 114–15; Charles Norton, *American Inventions and Improvements in Breech-loading Small Arms* (Boston, 1882), 295–96; V. Chebyshev, "Opisanie ustroistva metallicheskikh patronov, priniatykh dlia voennogo oruzhiia i posledovatel'nogo ikh usovershennstvovaniia," *O.S.* 2 (1871): 25; P. Lukin, "Novye obraztsy oruzhiia, zariazhaiushchegosia s kazni," *O.S.* 1 (1865), 12.

10. Alden Hatch, *Remington Arms in American History* (New York, 1956), 110, 142.

11. *Daily Telegraph*, 14 February 1877.

12. Tarassuk and Blair, *Complete Encyclopedia of Arms and Weapons*, 231–32; Smith, *Small Arms of the World*, 78.

13. Smith, *Small Arms of the World*, 100–102, 109–10; Tarassuk and Blair, *Complete Encyclopedia of Arms and Weapons*, 232–36; Hart, "Armed Forces," 306; G. Chinn, *The Machine Gun* (Washington, D.C., 1955), 180–81. Gatling contracted with Colt in 1867, and the Colt Company, although not involved in initial design or production, continued to make the gun for many years after. The scorpion metaphor comes from B. H. Liddell Hart, *The Remaking of Modern Armies* (Boston, 1928), 6.

14. Chinn, *The Machine Gun*, 179–80.

15. Norton, *American Inventions and Improvements* 207–9; W. B. Edwards, *The Story of Colt's Revolver: A Biography of Samuel Colt* (Harrisburg, Pa., 1953).

16. Joseph Rosa, *Col. Colt: London* (London, 1976), 16, 21; "Pistol," *Encyclopedia Britannica*, 11th ed., 21: 655; Schoen, *Rifled Infantry Arms*, 197–98. Adams's patent in 1851 prompted his employers, George and John Deane, to take him into partnership. Deane, Adams and Deane then undertook to manufacture the Adams revolver. Correspondence with Joe Rosa has helped me clear up some of the differences between Colt and the British revolvers.

17. "Firearms," *Colburn's United Service Magazine* 370 (September 1859): 51. See also Smith, *Small Arms of the World*, 160.
18. Russell Fries, "A Comparative Study of the British and American Arms Industries, 1790–1890," Ph.D. diss. (Johns Hopkins University, 1972), 171–72, 175, 184, 373; Rosa, *Col. Colt*, 129.
19. "Pistol," *Encyclopedia Britannica*, 11th ed., 21: 655; Norton, *American Inventions and Improvements*, 195; Roy G. Jinks, *History of Smith and Wesson* (North Hollywood, Calif., 1977), 1; Fries, "A Comparative Study," 174, 200. Two Soviet historians have claimed that the center-fire cartridge was first developed by the Russian gunsmith Vishnevskiii for his revolvers in the 1850s and later improved by Rakus-Sushchevskii and Pattsevich in 1866, but I have not been able to confirm this in any other source (N.I. Gnatkovskii and P. A. Shorin, *Istoriia razvitiia otechestvennogo strelkovogo oruzhiia* [Moscow, 1959], 105–108).
20. Daniel R. Beaver, "Cultural Change, Technological Development and the Conduct of War in the Seventeenth Century," in Russel F. Weigley, ed., *New Dimensions in Military History* (San Rafael, Calif., 1975), 79–80; Jay Luvaas, *The Military Legacy of the Civil War: The European Inheritance* (Chicago, 1959), 2–4.
21. T. Wintringham, *The Story of Weapons and Tactics from Troy to Stalingrad* (London, 1943), 136. William McNeil, *The Pursuit of Power: Technology, Armed Force and Society since A.D. 1000* (Chicago, 1982), 236.
22. Ernest R. Dupuy and Trevor Dupuy, *Encyclopedia of Military History* (London, 1977), 733–34, 820–24; Liddell Hart, *Remaking*, 49; Newman, *Tools*, 46; Michael E. Howard, *The Franco-Prussian War* (New York, 1961), 5–6, 24, 35; Cyril B. Falls, *A Hundred Years of War* (London, 1953), 68, 78; Archer Jones, *The Art of War in the Western World* (Urbana and Chicago, 1987), 400; Luvaas, *Military Legacy*, 2, 4, 46, 108, 111; McNeil, *The Pursuit of Power*, 252; Showalter, *Railroads and Rifles*, 82, 110–12, 118.
23. "The Military Power of Russia," *Edinburgh Review* (January 1878), 209.
24. Showalter, *Railroads and Rifles*, 83–85, 93–94, 107. I am also grateful to Jacob Kipp for his comments on this issue.
25. Both quoted in Smith, *Harpers Ferry Armory*, 156, 217.
26. Howard, *The Franco-Prussian War* 5, 35; Showalter, *Railroads and Rifles*, 103, 109, 123–24, 215–16.
27. J. F. C. Fuller, *A Military History of the Western World*, 3 vols. (New York, 1956), 3: 145; M. E. Howard, "The Armed Forces," *New Cambridge Modern History*, 11: 208–9.
28. John Keegan, *The Face of Battle* (London, 1976), 229–30.
29. "The Firearms Manufacture," 148.
30. Hounshell, *From the American System*, 15–25, 35, 46–50; Deyrup, *Arms Makers*, 3-5; Paul J. Uselding, "Studies of Technology in Economic History," in Robert E. Gallman, ed., *Recent Developments in the study of Business Economic History: Essays in Memory of Herman E. Krooss* (Greenwich, Conn., 1977), 168.

31. Claude Gaier-Lhoest, *Four Centuries of Liège Gunmaking* (London, 1976), 163. See also United States Senate, 36th Congress, 1st session, Executive Document no. 59, *Report on the Art of War in Europe in 1854, 1855 and 1856 by Major Richard Delafield, Corps of Engineers* (Washington, D.C., 1960), xv (hereafter cited as Delafield, *Report*).

32. Nathan Rosenberg, *The American System of Manufactures* (Edinburgh, 1969), 30–42; Showalter, *Railroads and Rifles*, 92, 239.

33. Showalter, *Railroads and Rifles*, 92; Beaver, "Cultural Change," 79.

34. Beaver, "Cultural Change," 79.

35. Deyrup, *Arms Makers*, 36–37; Fries, "A Comparative Study," 4. Dependence on foreign sources even included dependence on foreign and semifinished products. English iron, and later Sheffield steel, dominated the American market before 1850. See Geoffrey Tweedale, "Sheffield Steel and America: Aspects of the Atlantic Migration of Special Steelmaking Technology, 1850–1930," *Business History* 25, 3 (November 1983); 226.

36. Smith, *The Harpers Ferry Armory*, 189. For a study of the decline of laissez-faire and of the increasing state role in the British arms industry, see Clive Trebilcock, "War and the Failure of Industrial Mobilization, 1899–1914," in J.M. Winter, ed., *War and Economic Development: Essays in Memory of David Joslin* (Cambridge, 1975): 139–64.

37. Smith, *The Harpers Ferry Armory*, 324–25; Fries, "A Comparative Study," 6–7; Rosenberg, *The American System*, 68; Lee Kennett and James LaVerne Anderson, *The Gun in America* (Westport, Conn., 1975).

38. Nathan Rosenberg, "Technological Change in the Machine-Tool Industry, 1840–1910," *Journal of Economic History* 23 (December 1963): 414–43; idem, *Technology and American Economic Growth*, 103–5, 122–24; and idem, *Perspectives on Technology* (Cambridge, 1976), 19–20. See also Deyrup, *Arms Makers*, 3–5, and Smith, *The Harpers Ferry Armory*, 250–51.

39. Rosenberg, "Technological Interdependence," 46.

40. This is the thesis of Hounshell, *From the American System*, but the concept may be found in all the works cited in notes 37–39.

41. Smith, *The Harpers Ferry Armory*, 283, 324–25, 329.

42. Crouzet, "Recherches sur la production d'armements en France, 1815–1913," in *Conjuncture économique, structures sociales: Hommage à Ernest Labrousse* (Paris, 1974), 315.

43. Charles H. Fitch, *Report on the Manufacture of Fire-Arms and Ammunition*, Extra Census Bulletin (Washington, D.C., 1882), 6–7; Rosenberg, *Technology and American Economic Growth*, 93.

44. See, for example, Smith, *Harpers Ferry Armory*, 148–51.

45. Ibid., 81–84.

46. Ibid., 64–67, 272–73; Uselding, "An Early Chapter," 71–72.

47. Fitch, *Report*, passim; Rosenberg, *The American System*, passim; Robert S. Woodbury, *Studies in the History of Machine Tools* (Cambridge, Mass., 1972), passim.

48. Maurice Daumas, ed., *A History of Technology and Invention: Progress Through the Ages*, Eileen B. Hennessy, trans., 3 vols. (New York, 1979),

3: 105–6. See also Joseph Singer, *A History of Technology*, 7 vols. (Oxford, 1954–1978), 4: 426–27, 438–39.

49. Showalter, *Railroads and Rifles*, 81–82.

50. Fitch, *Report*, 5, 6, 13, 15, 26; Rosenberg, *The American System*, 60, 71; Rosenberg, "Technological Change in the Machine-Tool Industry: 414–43. Rosenberg makes the point also in *Technology and American Economic Growth*, 103–5, 122–24, and in *Perspectives on Technology*, 19–20. See also Deyrup, *Arms Makers*, 3–5. According to Mordecai, rifle balls for the Minié rifle had to be made by pressure in dies rather than by casting in molds (*Report*, 173–74). Paul Uselding notes that the process of metal removal by milling was developed in the United States before 1818. By the 1830s milling machines were a common adjunct to many American metalworking establishments. By contrast, milling machines were not widely used in Britain until the 1890s. See "Studies of Technology," 171. H. J. Habbakuk, *American and British Technology in the Nineteenth Century* (Cambridge, 1962), provides a useful introduction to the comparison of the two countries. Roderick Floud, *The British Machine-Tool Industry* (Cambridge, 1976), gives a more detailed survey of British machines.

51. See in particular the testimony of Josiah Whitworth in Rosenberg, *The American System*, 23–29, 48–49; Nathan Rosenberg, *Perspectives on Technology* (Cambridge, 1976), especially chapter 2; Hounshell, *From the American System*; Nathan Rosenberg, "Why in America?" in Otto Mayr and Robert C. Post, eds., *Yankee Enterprise: The Rise of the American System of Manufactures* (Washington, D.C., 1981), 49–62; Deyrup, *Arms Makers*; Vera Shlakman, *Economic History of a Factory Town: A Study of Chicopee, Massachusetts* (Northampton, Mass., 1935).

52. Fitch, *Report*, 7.

53. Cited in Hounshell, *From the American System*, 25–26. See also Rosenberg, *Technology and American Economic Growth*, 169; W. F. Durfee, "The History and Modern Development of the Art of Interchangeable Construction in Mechanism," *Transactions of the American Society of Mechanical Engineers* 14 (1893): 1225–57; idem, "The First Systematic Attempt at Interchangeability in Firearms," *Cassier's Magazine* 5 (1893– 1894): 469–77.

54. *The Times* (15 June 1857), 12.

55. Smith, *Harpers Ferry Armory*, 325.

56. Rosenberg, "Why in America?" 52-59; idem, *Technology and American Economic Growth*, 44.

57. Quoted in Hounshell, *From the American System*, 21.

58. Uselding, "Studies of Technology," 168.

59. Hounshell, *From the American System*, 41; Fries, "A Comparative Study," 44–47; Smith, *Harpers Ferry Armory*, 226–28, 249; Rosenberg, *Technology and American Economic Growth*, 104–5.

60. Rosenberg, *Technology and American Economic Growth*, 90. As Hounshell points out (*From the American System*, 15–25), the term was never precisely defined in the nineteenth century.

61. Singer, *History of Technology*, 4: 438

62. Kennett and Anderson, *The Gun in America,* 87–88; Deyrup, *Arms Makers,* 11; Fries, "A Comparative Study," 16–17, 21, 26–27; Hounshell, *From the American System,* 28–32; Edwin A. Battison, "Searches for Better Manufacturing Methods," *Tools and Technology* 3, 4 (Winter 1979): 13–18.

63. Cited in Hounshell, *From the American System,* 28.

64. Fries, "A Comparative Study," 41; Smith, *Harpers Ferry Armory,* 109–10, 219–49. Smith (p. 189) notes that government contracts provided much needed working capital in the form of money advances to private arms makers as well.

65. Rosenberg, *Technology and American Economic Growth,* 91.

66. *Report from the Select Committee on Small Arms,* Q. 1514, 1516 (Parliamentary Papers) (London, 1854): 116.

67. Samuel Colt, "On the Application of Machinery to the Manufacture of Rotating Chambered Breech Fire-Arms, and the Peculiarities of those Arms," *Minutes of the Proceedings of the Institute of Civil Engineers* 11 (1851–1852): 46.

68. *Report from the Select Committee on Small Arms,* Q. 1116, 1120–21, 1166, 87, 92; Hounshell, *From the American System,* 19, 21, 23, 49.

69. Howard, "Interchangeable Parts," 639.

70. Rosenberg, *Technology and American Economic Growth,* 91, 95; Smith, *Harpers Ferry Armory,* 283, 324–25.

71. Smith, *Harpers Ferry Armory,* 325.

72. Fitch, *Report,* 4–5, 24. Emphasis added.

73. David Granick, *Soviet Metal-Fabricating and Economic Development: Practice versus Policy,* 37.

74. Gene Silvero Cesari, "American Arms-Making Machine Tool Development, 1798–1855," Ph.D. diss. (University of Pennsylvania, 1970), 228–29; John Buttrick, "The Inside Contract System," *Journal of Economic History* 12, 3 (Summer 1952): 205–21. Hounshell states that inside contracting was "a major stimulus to nineteenth-century manufacturing technology" (*From the American System,* 49). See also Alfred Chandler, "The American System and Modern Management," in Mayr and Posts, eds., *Yankee Enterprise,* 156.

75. "The Firearms Manufactures," 148.

76. Smith, *The Harpers Ferry Armory,* 64–67, 335. See the seminal article by E. P. Thompson, "Time, Work-Discipline and Industrial Capitalism," *Past and Present* 38 (1967): 56–97; Eric Hobsbawm, *Labouring Men* (London, 1964); Herbert Gutman, "Work, Culture and Society in Industrializing America," *American Historical Review* 78 (1973): 531–87; Joan Wallach Scott, *The Glassworkers of Carmaux: French Craftsmen and Political Action in a Nineteenth-century City* (Cambridge, Mass., 1974).

77. Andrew Ure, *The Philosophy of Manufactures* (London, 1835), 16.

78. Uselding, "An Early Chapter," 73, 79–80; Fries, "A Comparative Study," 28–29.

79. Smith, *Harpers Ferry Armory,* 328.

80. Quoted in Blackmore, "Colt's London Armoury," in S. B. Saul, ed., *Tech-*

nological Change: U.S. and Great Britain in the Nineteenth Century (London, 1970), 182.

81. Samuel Smiles, *James Nasmyth: Engineer* (New York, 1883), 319.
82. *Report from the Select Committee on Small Arms*, Q. 1367, 108.
83. *Report from the Select Committee on Small Arms*, Q. 1367, 1441, 113.
84. Falls, *A Hundred Years of Warfare*, 18; James Everett Katz, ed., *Arms Production in Developing Countries: An Analysis of Decision Making* (Lexington, Mass. and Toronto, 1984), 5–7. This generalization applies equally to today's advanced countries at the moment when they were developing—such as the United States in the 1790s.
85. Crouzet, "Recherches," 307–9. The gradual privatization in France beginning in the 1870s occurred in the production of artillery pieces and ammunition; military small arms remained a government monoploy.
86. See Gary K. Bertsch, "Technology Transfers and Technology Controls: A Synthesis of the Western-Soviet Relationship," in Ronald Amann and Julian Cooper, eds., *Technical Progress and Soviet Economic Development* (Oxford, 1986), 116, 121; Josef C. Brada, "Soviet-Western Trade and Technology Transfer: An Economic Overview," in Bruce Parrott, ed., *Trade, Technology, and Soviet-American Relations* (Bloomington, Ind., 1985), 22–23; Mira Wilkins, "The Role of Private Business in the International Diffusion of Technology," *Journal of Economic History*, 34 (March 1974): 8–35; Shannon R. Brown, "The Transfer of Technology to China in the 19th Century," *Journal of Economic History* 39 (1979): 181–82; Katz, *Arms Production*, 43.
87. Katz, *Arms Production*, 8–9.
88. Brown, "Transfer," 183. Brown concluded that as an institutional form for modern enterprise the *kuan-to shang-pan* (literally, official supervision and merchant management) system attracted little private investment, was corrupt and inefficient, and was generally not successful. The comparable Russian hybrid fared little better. (See Chapter 6.)
89. Ibid., 197.
90. David Granick, *Soviet Metal-Fabricating*, 44, 47. For difficulties in a different setting, see Shannon R. Brown, "The Ewo Filature: A Study in Transfer of Technology to China in the 19th Century," *T. C.* 20, 3 (July 1979): 550—68; and idem, "The Transfer of Technology to China."
91. Mira Wilkins, *The Emergence of the Multi-national Corporation: American Business Abroad from the Colonial Era to 1914* (Cambridge, Mass., 1970), 29, 37, 66, 75–76.
92. Mordecai, *Report*, 162.
93. Quoted in "The Firearms Manufacture," 148; Fries, "A Comparative Study," 160, 260, 391.
94. A. Fon-der-Khoven, "Kratkie svedeniia o proizvodstve laboratornykh rabot, fabrikatsii metallicheskikh patronov i merakh, prinimaemykh v pravitel'stvennykh zavodakh SShA, dlia umen'sheniia poter' v liudiakh, pri vzryvakh i pozharakh," *O.S.* 3 (1881): 13.
95. Colt biographies and company histories, none of them scholarly, are: W. B. Edwards, *The Story of Colt's Revolver: Biography of Samuel Colt* (Harris-

burg, 1953); Charles T. Haven and Frank A. Belden, *A History of the Colt Revolver* (New York, 1940); James L. Mitchell, *Colt, the Arms, the Man, the Company* (Harrisburg, Pa., 1959); Jack Rohan, *Yankee Arms Maker* (New York, 1935); Martin Rywell, *Samuel Colt, A Man and an Epoch* (Harriman, Tenn., 1952); Colt Patent Fire Arms Manufacturing Company, *Colt Patent Fire Arms Manufacturing Company: A Century of Achievement, 1836–1936* (Hartford, Conn., 1937); and H. Barnard, *Armsmear: The Armory of Samuel Colt* (Boston, 1866).

96. Deyrup, *Arms Makers,* 220–22; Norton, *American Inventions and Improvements,* 207.

97. "Colt and His Revolvers," *Colburn's United Service Magazine* (1854): 118.

98. Samuel Colt to Elisha Colt, 18 July 1849, Vienna, Box 42, CPFAM Records, RG, 103, CSL.

99. All biographies of Colt note the importance of the Crystal Palace exhibit. The Imperial Exhibition at Paris sixteen years later helped Remington in the same way. See Hatch, *Remington Arms,* 144. Although accounts of the more famous international exhibitions exist, the history of the phenomenon has yet to be written. See Guy Stanton Ford, "International Exhibitions," *Encyclopedia of the Social Sciences* (New York, 1933), 6: 23–27; Great Exhibition of the Works of the Industry of All Nations, *Official Description and Illustrated Catalogue,* 3 vols. (London, 1851); David Burg, *Chicago's White City of 1893* (Lexington, Ky., 1976).

100. "Colt and His Revolvers," 120.

101. Colt to Kossuth, 20 March 1853, London, Samuel Colt Papers, Wadsworth Antheneum.

102. Last Will and Testament of Elizabeth Colt (1901), Colt Box, Folder Pre–1910, Samuel Colt Papers, Wadsworth Atheneum. See also Barnard, *Armsmear,* 307. Among the objects causing Colt's pockets to bulge were snuff boxes from Turkey and Russia, diamond rings from the tsars, a copy of the Koran, and, for good measure, two icons.

103. Colt to Prince Murat, Box 15 ("1851-56"), Samuel Colt Papers, CHS. The letter is signed "With the highest respect Sir—Your Highness [most dutiful—crossed out] Most [penciled in] Devoted Servaent [misspelled]." In the 1860s Remington followed a similar pattern in gaining acceptance at foreign courts. See Hatch, *Remington Arms,* 143.

104. Joe Rosa, *Col. Colt,* 28; Thomas Anquetil, *Notice sur les pistolets tournants et roulents, dits revolvers, ou leur passé, leur présent, leur avenir, suivié des principes generaux sur le tir de ces armes* (Paris, 1854), 16–17.

105. Colt, "On the Application of Machinery," 46. See also notice in *The Times,* 28 November 1851, 3.

106. Colt's testimony may be found in *Report from the Select Committee on Small Arms.* Much of the material on Colt in England is taken from Rosa, *Col. Colt,* especially pages 66, 70, 74. See also Rosenberg, *The American System,* 43–51, and John Rigley, "The Manufacturing of Small Arms," *Proceedings of the Institute of Civil Engineers,* vol. 111 (1893): 131.

107. Wilkins, *Emergence*, 30.
108. "Colt and His Revolvers," 121. See also Rosa, *Col. Colt*, 27–29, 37–43, 54–57.
109. *Colt's Patent Fire Arms*, 14.
110. Deyrup, *Arms Makers*, 124
111. See, for example, V. Bestuzhev-Riumin, "Ruchnoe oruzhie na parizhskoi mezhdunarodnoi vystavke," *O.S.* 4 (1867): 35–68; A. Fon-der-Khoven, "Zametki o venskoi vsemirnoi vystavke 1873 g.," *O.S.*, 5 installments (1873–1874); and *Otchet po venskoi vsemirnoi vystavke 1873 g. v voenno-tekhnicheskom otnoshenii*, 2 vols. (St. Petersburg, 1874, 1879). To take another example, demand for their revolvers in the United States had been such that until 1867 Smith and Wesson made no effort to sell them abroad. At the Paris Exposition, a case of their various models was exhibited, which at once attracted attention, and from that time a demand arose that constantly increased, resulting in large shipments to Japan, China, England, Russia, France, Spain, Peru, Chile, Brazil, and Cuba. (Norton, *American Inventions and Improvements*, 195.)
112. Norton, *American Inventions and Improvements*, 306.
113. V. Buniakovskii, "Ustroistvo Kol'tovskogo oruzheinogo zavoda," *O.S.* 4 (1869): 55–56; see also N. Alekseev, "Metallicheskii malokalibernyi patron," *O.S.* 1 (1877): 35.
114. *Daily Telegraph*, 14 February 1877.
115. Rosenberg, *Technology and American Economic Growth*, 90; Joseph Wickham Roe, "Interchangeable Manufacturing," reprint of address given before the American Society of Mechanical Engineers, Folder "R," Box 8, CPFAM Records, RG 103, CSL; Norton, *American Inventions and Improvements*, 321.

Chapter 3. Small Arms in Pre-Reform Russia

1. Regarding museums, foreigners were usually impressed. According to the Englishman Greener, "the arms museums of Russia are without equal for completeness and diverse system. . . . [They] contain more devices in arms mechanisms than would seem conceivable." See Major General Francis V. Greener, *The Russian Army and its Campaigns in Turkey, 1877–1878*, 2 vols. (London, 1879), 1: 76.
2. Mordecai, *Report*, 157.
3. Pintner, "The Burden of Defense," 232–34, 241.
4. The contemporary sources on Russian arms are the following: N. P. Pototskii and V. Shkliarevich, *Sovremennoe ruchnoe oruzhie, ego svoistra, ustroistvo i upotreblenie* (St. Petersburg, 1873), 335–85; *Sbornik noveishikh svedenii o ruchnom ognestrel'nom oruzhii dlia gg. pekhotnykh i kavaleriiskikh ofitserov russkoi armii* (St. Petersburg, 1857), 32; Schoen, *Pekhotnoe nareznoe ruzh'e*, 130, 208; and Mordecai, *Report*, 157. The opinion of the needle gun is cited from Col. A. Korostovtsev, "Obzor issledovaniia proizvedennykh u nas nad noveishimi sistemami ruchnogo ognestrel'nogo

oruzhiia," *A. Zh.* 1 (1854): 3; and A. M. Zaionchkovskii, "Vzgliad N. N. Murav'eva na sostoianie nashei armii i sobstvennoruchnye zamechaniia Imp. Nikolaia Pavlovicha," *V.S.* 2 (1903) 491. The secondary sources covering early nineteenth-century Russian arms include Curtiss, *The Russian Army,* especially 123–27; Fedorov, *Evoliutsiia;* and Mavrodin and Mavrodin, *Iz istorii,* 19–24.

5. Pintner, "Burden of Defense," 233, 237, 242–43. See also Beyrau, *Militär und Gesellschaft.*

6. "O sberezhenii ognestrel'nogo oruzhiia," *V.S.* 1 (1859): 178–79.

7. In the secondary sources, see Zaionchkovskii, *Voennye reformy,* 22; and Curtiss, *The Russian Army,* 123–25. The contemporary sources will be cited below.

8. Zaionchkovskii, *Voennye reformy,* 25.

9. Bushnell, *Mutiny amid Repression,* especially chapter 1.

10. A. V. Fedorov, *Vooruzhenie russkoi armii v Krymskuiu kampaniiu* (St. Petersburg, 1904), 121.

11. "Tsirkuliar Inspektorskogo Departamenta o tom, chto glinianye puli portiat stvoly," *Sbornik noveishikh svedenii,* 51–52. "O sberezhenii," 178–82. Even after the Crimean War, money for repairs was still controlled by the regulations of 1810. See Curtiss, *The Russian Army,* 121–26, for more on the same subject.

12. "O sberezhenii," 183.

13. Ibid., 185–86.

14. Ostroverkhov and Larionov, "Kurs o ruchnom ognestrel'nom oruzhii," *V.S.* 7 (1859): 176; M. I. Dragomirov, "Vliianie rasprostraneniia nareznogo oruzhiia na vospitanii i taktiku voisk," *O.S.* 1 (1861): 37; S. Vorob'ev, *Novoe ruchnoe ognestrel'noe oruzhie evropeiskikh armii* (St. Petersburg, 1864), 17.

15. More on tactical doctrine may be found in Major-General George B. McClellan, *The Armies of Europe: Comprising Descriptions in Detail of the Military Systems of England, France, Russia, Prussia, Austria, and Sardinia* (Philadelphia, 1861), 116–99, 211–94; Strokov, *Istoriia voennogo iskusstva;* Meshcheriakov, *Russkaia voennaia mysl';* Menning, *Bayonets Before Bullets;* Von Wahlde, "Military Thought"; and Walter Pintner, "Russian Military Thought: The Western Model and the Shadow of Suvorov," in Peter Paret, ed. *Makers of Modern Strategy from Machiavelli to the Nuclear Age* (Princeton, 1986): 354–75.

16. Curtiss, *The Russian Army,* 117, 120.

17. Fedorov, *Vooruzhenie* (1904), 121.

18. Ibid., 5.

19. Curtiss, *The Russian Army,* 120–21.

20. Bushnell, *Mutiny amid Repression,* 11–23; "O sberezhenii," 172–73.

21. Curtiss, *The Russian Army,* 120–21; Norton, *American Inventions and Improvements,* 301.

22. A. M. Zaionchkovskii, "Vzgliad," 12–15; Fedorov, *Evoliutsiia,* 6–7; Curtiss, *The Russian Army,* 121–23; Sir Lumley Graham, "The Russian Army

in 1882," *Journal of the Royal United Service Institute*, 4 installments (1883–1884): 482.

23. *The Times*, 17 December 1855, 83; V. G. Fedorov, *Vooruzhenie russkoi armii XIX v.* (St. Petersburg, 1911), 19, 22; *Fraser's Magazine* 50 (September 1854): 367.

24. Zaionchkovskii, *Voennye reformy*, 25; Beyrau, *Militär und Gesellschaft*, 100; Showalter, *Railroads and Rifles*, 99.

25. Fedorov, *Vooruzhenie* (1904), 8, 145; Fedorov, *Evoliutsiia*, 6; Mavrodin and Mavrodin, *Iz istorii*, 27–29.

26. Thomas Hart Seymour to Department of State, no. 45, 11 August 1855, "Despatches from U.S. Ministers to Russia," roll 16, Microcopy no. 35, RG 59, NA.

27. Cyril Falls, *A Hundred Years of Warfare* 30; Showalter, *Railroads and Rifles*, 99.

28. Menning, *Bayonets Before Bullets*, 39.

29. V. G. Fedorov, *Vooruzhenie* (1904), 8; Fedorov, *Evoliutsiia*, 6; Liddell Hart, "Armed Forces," 322. Jay Luvaas reaches a similar conclusion (*Military Legacy*, 2). England's rapid deployment of the Enfield rifle with the Minié ball may have been, in part, a response to perceived Russian advantages in artillery. According to *The Times*, "The Russian field artillery appears fully as good, if not better than the British and is of much heavier metal. Cannot the power of this Russian field artillery to a certain degree be nullified?" (27 December 1854, 5).

30. "Otchet za trekhletnee upravlenie artilleriei Ego Imperatorskim Vysochestvom General'nym-Fel'dtseikhmeisterom, s 25–go ianvaria 1856 po 25-oe ianvaria 1859g," *A. Zh.* 5 (1860) 301–8; Fedorov, *Vooruzhenie* (1904), 26; *Istoriia Tul'skogo oruzheinogo zavoda, 1712–1972* (Tula, 1973), 449. Ironically, on the eve of the war the Russian government had requested, and had received, from the British Board of Ordnance a set of drawings of machinery invented by the Englishman John Anderson to produce the conoidal bullet for the Minié rifle. There is no evidence that this machinery was made in Russia in time to be of use during the Crimean War (Rosenberg, *The American System*, 84–85).

31. "Tsirkuliar Inspektorskogo departamenta Voennogo ministerstva," no. 3 (13 October 1855), published in *Sbornik noveishikh svedenii*, 196; Fedorov, *Vooruzhenie* (1904), 30–31.

32. *The Times*, 4 April 1857, 7.

33. V. Veshniakov, "Russkaia promyshlennost' i ee nuzhdy." *Vestnik Evropy* 11 (1870): 133.

34. Fedorov, *Vooruzhenie* (1904), 13–14. See also N.I. Gnatovskii and P. A. Shorin, *Istoriia razvitiia otechestvnnogo strelkovogo oruzhiia*, and P. A. Zaionchkovskii, "Perevooruzhenie russkoi armii v 60-70kh gg. XIX v."

35. Fedorov, *Vooruzhenie* (1904), 13–14; Curtiss, *The Russian Army*, 127.

36. M. Malkin, "K istorii russko-amerikanskikh otnoshenii vo vremia grazhdanskoi voiny v SShA," *Krasnyi arkhiv* 14 (1939): 97; Andrew D. White, *Autobiography* (New York, 1905), 1: 454; Thomas Hart Seymour

to Department of State, no. 3, 8 May 1854, "Despatches from U.S. Ministers to Russia," roll 16, Microcopy no. 35, RG 59, NA. Additional accounts, not in all cases scholarly, of the amicable relations between Russia and America may be found in Norman E. Saul, "Beverly C. Sanders and the Expansion of American Trade with Russia, 1853–1855," *Maryland Historical Magazine* 67, 2 (Summer 1972): 160; Erwin Holze, *Russland und Amerika: Aufbruch und Begegnung Zweier Weltmachte* (Munich, 1953), 194–203; James Road Robinson, *A Kentuckian at the Court of the Tsars: The Ministry of Cassius Clay to Russia, 1861–62 and 1863–69* (Berea, Ky., 1935), 256–60; Frank A. Golder, "Russian-American Relations during the Crimean War," *American Historical Review* 31, 3 (April 1926): 466; Alexander Tarsaidze, *Czars and Presidents: The Forgotten Friendship* (New York, 1958).

37. *The Times,* 12 January 1855, 10. The article, reprinted from *The New York Herald,* defended American friendliness to Russia during the Crimean War.

38. "American Genius and Enterprise," *Scientific American* 2 (September 1847): 397. Quoted in Leo Marx, *The Machine in the Garden* (New York, 1964), 205–6.

39. Pintner, "The Burden of Defense," 3. William L. Blackwell, *The Beginnings of Russian Industrialization, 1800–1860* (Princeton, 1968), 173–74. Official Russian trade statistics do not provide the overall size or the composition of Russian-American trade. Government purchases, as well as equipment, machines, and tools, were not listed. In addition, a large part of the trade was done through intermediaries. See Ministerstvo finansov, Departament vneshnei torgovli, *Gosudarstvennaia vneshniaia torgovlia v raznykh ee vidakh za 1854–64,* 10 vols. (St. Petersburg, 1855–1865). This problem is further discussed in Kirchner, *Studies,* 44–45, 66, 69.

40. Carstensen, *American Enterprise,* 39; Kirchner, "Russian Entrepreneurship," 87; Crisp, *Studies,* 11; F. S. Claxton to Department of State, 2 October 1857, Moscow, "Despatches from U.S. Consuls in Moscow, 1857–1906," roll 1, Microcopy no. 456, RG 59, NA.

41. G. M. Hutton to Department of State, 17 November 1856, "Despatches from U.S. Consuls in St. Petersburg, 1803–1906," rolls 5–6, Microcopy no. 81, NA.

42. *The Times,* 6 November 1855, 6; Kirchner, *Studies,* 140. Indeed, the British government protested that in supplying arms and ammunition to Russia, the United States was violating neutrality laws (see United States Congress, Senate, 34th Congress, 1st session, Senate Executive Documents, no. 27, 29, 34).

43. *Nile's National Register* 61 (2 October 1841): 80; Morskoe ministerstvo, *Istoricheskii obzor razvitiia i deiatel'nosti Morskogo ministerstva za 100 let ego sushchestvovaniia,* S. F. Ogorodnikov, comp. (St. Petersburg, 1902), 126–27; Thomas Hart Seymour to Department of State, no. 78, 28 May 1856, "Despatches from U.S. Ministers to Russia," roll 17, Microcopy no. 35, NA; *The Times,* 12 January 1855, 10, and 14 November 1857, 8; Phillip

K. Lundeberg, *Samuel Colt's Submarine Battery: The Secret and the Enigma*, Smithsonian Studies in History and Technology, no. 29 (Washington, D.C., 1974), 4–5, 19.

44. Lundeberg, *Submarine Battery*, 18, 25. Edwards, *The Story of Colt's Revolver*, 128, recounts the same incident. According to Colt historian R. L. Wilson, "among the known recipients of presentation arms from Colt in the Paterson period were . . . Czar Nicholas I of Russia." See R. L. Wilson, *Colt, An American Legend: The Official History of Colt Firearms from 1836 to the Present* (New York, 1985), 56, 50, 125–26, 132–36.

45. "O povtoritel'nom ognestrel'nom oruzhii ili revol'verakh," *A. Zh.* 5 (1855): 1–34. One year later this article was reprinted in *Morskoi sbornik* 2 (1856): 426–59. Gorlov used the French version of Colt's brochure, *Armes à feu à culasse tournante du colonel Saumel Colt* (Brussels, 1854), and patent descriptions.

46. K. Kostenkov, *Opisanie revol'verov i pravila obrashchat'sia s nimi* (St. Petersburg, 1855), 16.

47. *Morskoi sbornik* 16, 6 (June 1855): 163; Konstantinov, "Posledovatel'nye usovershenstvovaniia ruchnogo ognestrel'nogo oruzhiia," *Morskoi sbornik* 5 (1855): 48–49.

48. Fretwell to Colt, 19 April 1853, London, Folder "Firth and Sons," CPFAM Records, RG 103, CSL. Firth had begun to develop a market in Russia for its files and tool steel. See Fred Carstensen, "Foreign Participation in Russian Economic Life: Notes on British Enterprise, 1865–1914," in Gregory Guroff and Fred Carstensen, eds., *Entrepreneurship in Imperial Russia and the Soviet Union* (Princeton, 1983), 140–59.

49. Konstantinov, "Posledovatel'nye usovershenstvovaniia," 48–49. See also Mikhail Terent'ev, "Vzgliad na istorii i sovremennoe sostoianie povtoritel'nogo oruzhiia, ili revol'verov," *V.S.* 12 (1860): 215–68; Mikhail Kovalevskii, "Sovershenstvovaniia ruchnogo ognestrel'nogo oruzhiia v techenie XIX v.," *O.S.* 2 (1903): 75.

50. Sainthill to Colt, 27 October 1853, 12 November 1853, and 2 March 1854, Brussels, Box 13, CPFAM Records, RG 103, CSL; White, *Autobiography*, 1: 454; V. G. Fedorov, *Vooruzhenie* (1904), 19.

51. U.S. Legation, St. Petersburg, 29 October 1854, Documents Received, 1853–58, vol. 4342, "Department of State Foreign Service Posts," RG 84, NA; and Alden to Colt, CPFAM Records, RG 103, CSL; Leonid Tarassuk and R. L. Wilson, "The Russian Colts," *The Arms Gazette* (August-October 1976). Wilson and Tarassuk claim that Nicholas's trip was "secret." Other than the understandable desire to be discreet regarding weapons negotiations, it seems doubtful that the emperor traveled in secrecy. V. V. Mavrodin also describes the Colt revolvers in the Ermitage collection ("Revol'very Tul'skikh oruzheinikov," *Soobshcheniia Ermitazha* 40 [1975]: 36). During all three trips the Russians in return gave Colt many gifts, including snuff boxes ornamented with diamonds and the tsar's signature and "twelve kettles of different sizes and qualities of peculiar make in Russia for heating water, coffee, tea, etc etc and known in St. Petersburg by the name of

'samivar' (sic)." Colt received similar gifts from the King of Siam, Prince Albert, the Sultan, and other members of royalty all over the world, (Colt to Joslin, 15 October 1856, St. Petersburg, Box 15, Samuel Colt Papers, CHS).

52. Catacazy to Hamilton Fish, 11 April 1870, Washington, "Notes from the Russian Legation in the U.S. to the Department of State, 1809–1906," roll 4, Microcopy no. 39, NA; Fedorov, *Vooruzhenie* (1904), 19–20. Alexander Tarsaidze, with no documentation, places this visit "shortly before the Crimean War ("Berdanka," *Russian Review* 4, 4 [January 1950]: 32). The daily newspaper of the Russian army noted that Lilienfel'd, who devised his own carbine system and became the director of the Sestroretsk small arms factory, successfully continued his technical education in America, although there is no indication when or where this occurred. See L. P., "Kavalerskii pistolet polkovnika Lilienfel'da," *Russkii invalid* 109 (19 May 1870): 4.

53. Captain Jervis (Office of Ordnance) to Foreign Office, 16 July 1855, New York, Foreign Office Records, FO 65/464, PRO.

54. Colt to Sergeant, 20 April 1855, Hartford, Box 3, Folder "Colt," CPFAM Records, RG 103 CSL; "Postanovlenie i rasporiazhenie pravitel'stva," Tsirkuliary Inspektorskogo departamenta, no. 95 (9 April 1856): "Ob'iavlenie Artilleriiskogo departamenta Morskogo ministerstva," *Morskoi sbornik* 23, 8 (June 1856): liii; *The Times*, 19 December 1854 and 19 December 1855, 8; Fedorov, *Vooruzhenie* (1904), 15.

55. Joslin to Stoeckl, 29 May 1855, Hartford, Box 13, Cylinder 46, CPFAM Records, RG 103, CSL. The arms seizure threatened to damage Colt's reputation and his business with the British, and company archives contain several letters on "damage control." Sergeant to Colt, 24 August 1855, London, Box 9, Folder S, CPFAM Records, RG 103 CSL, clipping from *The Times* attached (no date, but refers to a letter of 11 August); Sainthill to Sergeant, 17 August 1855, Brussels, Box 13, Cylinder 56, CPFAM Records, RG 103 CSL. See also Joseph Rosa, "The KM Colts—A Few More Facts," *The Gun Report* (February 1985): 48–50; and Rosa, *Col. Colt*, 96.

56. Fedorov, *Vooruzhenie* (1904), 20, citing a letter from Colt to the Artillery Department. No copy of the letter appears in the Colt Papers or in the CPFAM Records. Fedorov bases his account on unpublished material of the Main Artillery Administration, Office of the Quartermaster General of Ordnance, no. 2323, for 1854. According to the Archival Guidebook, the correspondence concerning rifle orders in America are in the Artillery Historical Museum (AIM), now named the Military-Historical Museum of the Artillery, Corps of Engineers, and Sappers, hereafter cited as (AVIMAIVS), in Leningrad (See I. P. Ermoshin, ed., *Putevoditel' po Artilleriiskomu istoricheskomu arkhivu* [Leningrad, 1957], 86).

57. Samuel Colt to Joslin, 3 September 1856, Moscow, Box 15, Samuel Colt Papers, CHS. Colt added, "Time is absorbed without any apparent advantage gained in a business point of view. I have now been in the country upwards of three weeks and as yet the only thing we have been able to

accomplish is to set a man . . . at work repairing the broken or defective arms of the last 3,000 delivered by Mr. Caesar." (Samuel Colt to James Colt, 21 September 1855, Hartford, Box 42, Folder "James Colt, 1850–55," CPFAM Records, RG 103, CLS). At the same time, the Russian government was negotiating with other Americans, with not altogether favorable results. According to Fedorov, in 1855 a merchant by the name of Peters offered to provide the Russian government with 50,000 to 150,000 rifles at 15 rubles apiece at an undisclosed border point. Russia agreed to buy 50,000 but refused to make an advance payment. When Peters later insisted on an advance of 100,000 rubles, claiming that commercial agents in New York required payment in advance for shipping, "it became obvious that the Americans, not having the means to fill the order, were counting on the constraints of the Russian government for their own selfish ends." To Fedorov, writing even before the Soviet period, "The effort to order arms in America provides the most vivid example of the unethical behavior of arms makers who wanted to take advantage of Russia's awkward position" (*Vooruzhenie* [1904], 18).

58. Colt to Joslin, 2 August, 18 August, 3 September, 30 September, and 4 October 1856, St. Petersburg and Moscow, Box 15, Samuel Colt Papers, CHS.

59. The first indication of such sales to naval officers appears in *Morskoi sbornik* 15, 3 (1855): 292. According to Fedorov, *Vooruzhenie* (1904), 39, regimental commanders at Sevastopol had Colt revolvers. See also S. Zybin, M. Nekliudov, and M. Levitskii, *Oruzheinye zavody: Tul'skii, Sestroretskii, Izhevskii* (Kronstadt, 1898), 1. Mavrodin, "Revol'very Tul'skikh oruzheinikov," 34–36, cites unpublished material in AVIMAIVS. Descriptions of the Tula revolvers now in Soviet museums are in M. M. Denisova, *Russkoe oruzhie: kratkii opredelitel' russkogo boevogo oruzhiia* XI–XIX v. (Moscow, 1953), 121.

60. A. Gorlov, "O povtoritel'nom ognestrel'nom oruzhii ili revol'verakh," *Sbornik noveishikh svedenii*, 519. It is quite likely that the 1851 navy model made at the Izhevsk small arms factory and now owned by Al Weatherhead was one of these revolvers.

61. V. N. Ashurkov, "K istorii promyshlennogo perevorota v gosudarstvennom oruzheinom proizvodstve," in *Iz istorii Tuly i Tul'skogo kraia: Sbornik nauchnykh trudov* (Tula, 1983), 50; "Istoricheskii obzor Tul'skogo oruzheinogo zavoda i nastoiashchee ee polozhenie," *O.S.* 4 (1873): 5. The Soviet historian of the Izhevsk small arms factory elusively states that rifling machines "appeared" in the late 1850s. See A. A. Aleksandrov, *Izhevskii zavod: Nauchno-populiarnyi ocherk istorii zavoda, 1760–1917* (Izhevsk, 1957), 57–61.

62. Mordecai, *Report*, 159; Colt to Joslin, 2 August and 4 October 1856, Box 15, Samuel Colt Papers, CHS, refers to tools and machinery. See also *The Times* (11 December 1854), 10.

63. Ashurkov, "K istorii promyshlennogo perevorota v gosudarstvennom oruzheinom proizvodstve," 58.

64. Colt to Joslin, 2 August and 4 October 1856, Stettin and St. Petersburg, Box 15, Samuel Colt Papers, CHS; Colt to C. F. Dennet, 27 September 1859, Hartford, Box 13, CPFAM Records, CSL; Box 41, Folders "Miscellaneous," "Colt Collections of Firearms," CPFAM Records, CSL. It would appear that Colt made inquiries about his application for a patent (Colt to Caeser, 6 April 1857, Hartford, Box 15, Samuel Colt Papers, CHS).

65. "Jarvis letters," 5, 12, and 26 April 1858, Hartford, Samuel Colt Papers, CHS; *The Times* (6 January 1857), in Box 44, "Scrapbook, 1854–1860," CPFAM Records, RG 103, CSL.

66. "Perechen' zaniatii Oruzheinoi komissii," *Oruzheinyi sbornik* 2 (1861): 22; Vorob'ev, *Novoe oruzhie*, 125.

67. "Perechen' zaniatii Komiteta ob uluchshenii shtutserov i ruzhey," *A. Zh.* 1 (1857): 67–68; 2 (1857): 129; U.S. Legation, St. Petersburg, 8/20 May 1857, Documents Received, 1856–59, vol. 4344, "Department of State Foreign Service Posts," RG 84, NA.

68. "Perechen' zaniatii Oruzheinoi komissii," *A. Zh.* 9 (September 1861): 427–78. See also Vorob'ev, *Novoe oruzhie*, 125; Fedorov, *Vooruzhenie* (1904), 146; P. Lukin, "Kollektsiia ruchnogo ognestrel'nogo oruzhiia inostrannykh obraztsov, prinadlezhashchikhsia Oruzheinoi komissii," *O.S.* 4 (1862): 29. Such was also the conclusion of a manual of hunting guns. See K. Bol'dt, "Rukovodstvo k izucheniiu okhotnich'ego oruzhiia," *O.S.* 4 (1863): 84.

69. "Pravitel'stvennoe rasporiazhenie, no. 331 (14 September 1865)," *O.S.* 4 (1865): 2 (on prison guards); "Perechen' zaniatii Oruzheinoi komissii s pervogo ianvaria 1863 g.," *O.S.* 3 (1863). 13–15 (on officers of the Third Section); "Pravitel'stvennoe rasporiazhenie, no 81 (2 May 1865) and no. 187 (6 August 1866)," *O.S.* 2 (1865): 9; and no. 4 (1866), 10–11; F. S. Claxton to Department of State, 10 April 1858, Moscow, "Despatches from U.S. Consuls in Moscow, 1857-1906," roll 1, Microcopy no. 456, RG 59, NA; *Sbornik noveishikh svednenii* (1857), 560–63.

70. A. Gorlov, "Ob usovershenstvovanii pistoletov-revol'verov sistemy Adamsa," *A. Zh.* 1 (1856): 17–22.

71. "Perechen' zaniatii Oruzheinoi komissii," *O.S.* 4 (1862): 58. See also Lukin, "Kollektsiia ruchnogo ognestrel'nogo oruzhiia," 29; Ostroverkhov and Larionov, "Kurs, " 174–75.

72. Mavrodin, "Revol'very," 36.

73. White, *Autobiography*, 1: 454.

74. "Perechen' zaniatii Komiteta ob uluchshenii shtutserov i ruzhei," *A. Zh.* 1 (1857): 67-8; 2 (1857): 129. See also Aleksandrov, *Izhevskii zavod*, 61.

Chapter 4. Russian Small Arms Industry

1. N. O. S. Turner, F. G. E. Warren, and J. P. Nolan, *Tour of Artillery Officers in Russia* (London, 1867), 32. See also V. N. Zagoskin, "O tekhnicheskikh usloviiakh deshevogo proizvodstva metallicheskikh patronnykh gil'z," *O.S.* 3 (1875): 29–30.

2. A. Orfeev, "Istoriia Tul'skogo oruzheinogo zavoda," *O.S.* 3 (1903): 32; N. N. Kononova, "Rabochie oruzheinykh zavodov voennogo vedomstva v pervoi polovine XIX v.," *Uchenye Zapiski Leningradskogo gosudarstvennogo universiteta,* vyp. 32, no. 270 (1959), 121.

3. V. Chebyshev, "Po voprosu o novom administrativnom i tekhnicheskom ustroistve Tul'skogo oruzheinogo zavoda," *V.S.* 12 (1869): 249.

4. Gamel', *Opisanie*, x; F. Graf, "Oruzheinye zavody v Rossii," *V.S.* 9 (1861): 113–16; V. N. Ashurkov, *Gorod masterov* (Tula, 1958), 47-9; *Istoriia Tul'skogo oruzheinogo zavoda*, 24–31.

5. *Svod zakonov Rossiiskoi imperii*, 1857 edition, vol. 11, pt. 2: *Ustav o promyshlennosti*, article 36; article 66 in the 1887 edition; Graf, "Oruzheinye zavody," 117–21; M. Subbotkin, "Ob Izhevskom oruzheinom zavode," *O.S.* 2 (1863): 150–52.

6. Samuel H. Baron, "Entrepreneurs and Entrepreneurship in Sixteeth-and Seventeenth-Century Russia," in Guroff and Carstensen, *Entrepreneurship*, 56.

7. Tugan-Baranovsky, *The Russian Factory*, 93, 99; Kaser, "Russian Entrepreneurship," 440.

8. Graf, "Oruzheinye zavody," 114–21.

9. The following is based on Graf, "Oruzheinye zavody," 123–24; P. M. Maikov, "O proizvoditel'nykh silakh oruzheinykh zavodov Izhevskogo, Tul'skogo i Sestroretskogo," *A. Zh* 8, 9 (1861): 583–642; 1, 2 (1862): 38–77, 122–54; M. Subbotkin, "Ob Izhevskom oruzheinom zavod," *O.S.* 2 (1863): 150–78; V. N. Ashurkov, *Kuznitsa oruzhiia: Ocherki po istorii Tul'skogo oruzheinogo zavoda* (Tula, 1947). Despite the considerable technical and production changes of the 1870s, to be discussed in Chapter 6, the administrative structure was little changed at the close of the century. See *Pamiatnaia knizhka Tul'skoi gubernii na 1888 g.* (Tula, 1888), 33–34.

10. D. P. Strukov, *Glavnoe artilleriiskoe upravlenie: Istoricheskii ocherk*, in *Stoletie Voennogo Ministerstva*, vol. 6 (St. Petersburg, 1902).

11. Graf, "Oruzheinye zavody," 127.

12. Ibid.

13. "Rabochie oruzheinykh zavodov," 142; Orfeev, "Istoriia Sestroretskogo zavoda," 31. See also Beyrau, *Militär und Gesellschaft*, 99.

14. *Russia and the United States Correspondent*, no. 3, 30 August 1856, 2.

15. Graf, "Oruzheinye zavody," 131, 367, 379; Subbotkin, "Ob Izhevskom," 152; *Sestroretskii instrumental'nyi zavod: Ocherki, dokumenty, vospominaniia, 1721–1967* (Leningrad, 1968), 88; M. I. Gorbov, *Izhevskie oruzheiniki* (Izhevsk, 1963), 19; Aleksandrov, *Izhevskii zavod*, 65; V. A. Liapin, "Rabochie voennogo vedomstva vo vtoroi polovine XIX v." in L. V. Olkhovaia, ed., *Genezis i razvitie kapitalisticheskikh otnoshenii na Urale* (Sverdlovsk, 1980), 36; V. D. Zelentsov, "Otmena krepostnogo prava na Izhevskom zavode v 1866 g.," *Uchenye zapiski Gor'kovskogo pedagogicheskogo instituta*, vyp. 97 (1934), 74–75, 81–86.

16. "Artel'," *Entsiklopedicheskii slovar'*, 82 vols. (St. Petersburg: Brokgauz i Efron, 1890–1902), 3: 184–96: Kaser, "Russian Entrepreneurship," 428–30.

17. Aleksandrov *Izhevskii zavod,* 59; Ashurkov, *Kuznitsa oruzhiia,* 50; Graf, "Oruzheinye zavody," 125; Afremov *Istoricheskoe obozrenie,* 205, 236; P. Glebov, "Koe-chto o tul'skikh oruzheinikakh," *A. Zh.* 2 (1862): 176; "Tul'skii oruzheinyi zavod," 113; Gamel', *Opisanie,* 130–31.

18. Gamel', *Opisanie,* 61–62; Graf, "Oruzheinye zavody," 129–30, 369, 379; Glebov, "Koe-chto," 165–75; Kononova, "Rabochie oruzheinykh zavodov," 120; N. F. Firsanova, "K istorii sosloviia tul'skikh kazennykh oruzheinikov v period krizisa krepostnichestva," in V. V. Mavrodin, ed., *Rabochie oruzheinoi promyshlennosti* (Leningrad, 1976), 17.

19. Glebov, "Koe-chto," 161–78; Maikov, "O proizvoditel'nykh silakh," 59–61; Kononova, "Rabochie oruzheinykh zavodov," 125.

20. Glebov, "Koe-chto," 165–75; Graf, "Oruzheinye zavody," 129–30, 369, 379; Kononova, "Rabochie oruzheinykh zavodov," p. 120; N. F. Trutnev, "Tul'skaia oruzheinaia sloboda i kazennyi zavod v pervoi chetverti XVIII v." in V. N. Ashurkov, ed., *Iz istorii Tuly i Tul'skogo kraia* (Tula, 1983), 130; Britkin, *The Craftsmen of Tula,* 2; Liapin, "Rabochie voennogo vedomstva," 30–34.

21. Tugan-Baranovsky, *The Russian Factory,* 99.

22. Glebov, "Koe-chto," 161–62, 176.

23. Ibid., 171–72, 178; Maikov, "O proizvoditel'nykh silakh," 59.

24. *Istoriia rabochikh Leningrada, 1703–1965,* 2 vols. (Leningrad, 1972), 1: 81, 92.

25. Kononova, "Rabochie oruzheinykh zavodov," 129.

26. Ibid., 122–23; *Istoriia rabochikh Leningrada,* 1: 21.

27. A. Romanov, "Izhevskii oruzheinyi zavod: Mediko-topograficheskii ocherk," *Sbornik sochinenii po sudebnoi meditsine* 3 (1875): 20; A. Solov'ev, *V pamiat' stoletiia iubileia osnovaniia Izhevskogo oruzheinogo zavoda* (Izhevski, 1907), 13; *Istoriia rabochikh Leningrada,* 1: 81.

28. *Istoriia rabochikh Leningrada,* 1: 81; Glebov, "Koe-chto," 161–78; Maikov, "O proizvoditel'nykh silakh," 59–61.

29. At Izhevsk a four-year gunmaking school opened in 1870 and a trade school opened in 1877. See Aleksandrov, *Izhevskii zavod,* 111.

30. In addition to the instance of Tula gunmakers being transferred to Sestroretsk, in the early years of the Izhevsk armory, workers were sent to the Aleksandrovsk factory in St. Petersburg and to the Kamensk factory in Ekaterinburg to learn forging techniques. However, factory histories do not indicate that this practice was repeated. See Gorbov, *Izhevskie oruzheiniki,* 22.

31. Glebov, "Koe-chto," 182; *Istoriia rabochikh Leningrada,* 1: 104.

32. Glebov, "Koe-chto," 182; Orfeev, "Istoriia," 30–31. A Soviet study of Izhevsk points out that while barrel defects at Izhevsk frequently reached 70 percent, barrel defects at Liège, where the same Urals iron was used, averaged 7.5 percent (Liapin, "Rabochie voennogo vedomstva," 38).

33. *Istoriia rabochikh Leningrada,* 1: 104.

34. Graf, "Oruzheinye zavody," 128; Kononova, "Rabochie oruzheinykh zavodov," 131.

35. Graf, "Oruzheinye zavody," 129–30; Romanov, "Izhevskii oruzheinyi zavod," 20–21; Kononova, "Rabochie oruzheinykh zavodov," 120; Liapin, "Rabochie voennogo vedomstva," 34; *Istoriia rabochikh Leningrada,* 1:115. The Russian average was actually even lower (261 days), but the European average was much higher (305 days).

36. Tugan-Baranovsky, *The Russian Factory,* 93.

37. Cited in Glebov, "Koe-chto," 164.

38. "Ob Izhevskom oruzheinom zavod," 150; *Istoriia Tul'skogo oruzheinogo zavoda,* 8.

39. Glebov, "Koe-chto," 165, 176–78. It is not accidental that Glebov was writing in 1862.

40. Orfeev, "Istoriia Sestroretskogo oruzheinogo zavoda," 30–31.

41. Afremov, *Istoricheskoe obozrenie,* 240.

42. Herbert Barry, *Russian Metallurgical Works, Iron, Copper, and Gold, Concisely Described* (London, 1870), 41.

43. Glebov, "Koe-chto," 176.

44. Liapin, "Rabochie voennogo vedomstva, 32.

45. "Istoricheskii obzor Tul'skogo oruzheinogo zavoda i nastoiashchee ego polozhenie," *O.S.* 4 (1873): 6–9; Gamel', *Opisanie,* 87.

46. Robert Bremner, *Excursions in the Interior of Russia* (London, 1839), 2: 308–9.

47. Graf, "Oruzheinye zavody," 134–35; Gamel', *Opisanie,* x; Kononova, "Rabochie oruzheinykh zavodov," 135–36.

48. Graf, "Oruzheinye zavody," 372; Kononova, "Rabochie oruzheinykh zavodov," 135–36.

49. Glebov, "Koe-chto," 165–75; Maikov, "O proizvoditel'nykh silakh," 60–61. See also Graf, "Oruzheinye zavody," 372; and Kononova, "Rabochie oruzheinykh zavodov," 135–36.

50. Graf, "Oruzheinye zavody," 372–73.

51. "Ob Izhevskom zavode," 164; Gamel', *Opisanie,* 139–40, 145. On Sestroretsk, see Orfeev, "Istoriia Sestroretskogo oruzheinogo zavoda," 30.

52. Bremner, *Excursions,* 2: 310–11. Twenty years later a visiting U.S. military commission believed that "Russian arms were superior in finish to the others." See Mordecai, *Report,* 91.

53. Glebov, "Koe-chto," 173–74; Graf, "Oruzheinye zavody," 135–36.

54. Gamel', *Opisanie,* x; Trutnev, "Tul'skaia oruzheinaia sloboda" 118.

55. Quoted in Ashurkov, *Kuznitsa oruzhiia,* 51.

56. Ibid.

57. Ashurkov, "K istorii promyshlennogo perevorota," 54.

58. Gamel', *Opisanie,* x. Also quoted in F. N. Zagorskii, *A History of Metal Cutting Machines to the Middle of the Nineteenth Century,* trans. Edwin A. Battison (New Delhi, 1982), 231.

59. Tegoborski, *Commentaries,* 2: 110, 131.

60. Ministerstvo finansov, *The Industries of Russia* (St. Petersburg, 1893), 178–79; Ia. S. Rozenfel'd and K. I. Klimenko, *Istoriia mashinostroeniia SSSR* (Moscow, 1961), 25.

61. Gamel', *Opisanie*, i. Gamel', a collegiate counselor and doctor of medicine, was instructed to survey the manufactories of Tula province. I am grateful to Edwin Battison of the American Precision Museum for helping me track down a copy of this rare book at the Smithsonian libraries. Battison has written an introduction to a recent English translation of the study by Gamel', which was available only after my own manuscript had been completed. The translation was undertaken by Franklin Books Program, Inc., Cairo, and published for the Smithsonian Institution Libraries and the National Science Foundation. See Iosif Gamel', *Description of the Tula Weapon Factory in Regard to Historical and Technical Aspects* (New Delhi: Amerind Publishing Co., 1988). The manuscript collection of the Smithsonian Libraries holds a collection of papers and letters of Gamel'; regrettably, nothing in this collection pertains to the arms industry.

62. Gamel', *Opisanie*, xv; Afremov, *Istoricheskoe obozrenie*, 214–15.

63. Gamel', *Opisanie*, 117, 146.

64. Ibid., 131–32.

65. Britkin, *The Craftsmen of Tula*, 32, 38, 40. The director of the new Izhevsk plant bought English tools in St. Petersburg in 1808 (Solov'ev, *V pamiat'*, 15). Praise for Zakhava's mechanical genius can also be found in an 1818 description of the Tula factory. See "Tul'skii oruzheinyi zavod," 104–8.

66. Gamel', *Opisanie*, 73, 117, 146.

67. Ibid., 53–54, 200.

68. Ibid., viii, 201, 262. Gamel' likened Jones's contribution to the manufacture of small arms to that of the Scot Gascoyne in the field of artillery (p. viii).

69. *Sestroretskii instrumental'nyi zavod*, 25–27, 48.

70. Aleksandrov, *Izhevskii zavod*, 50. This study was one of the volumes commissioned for the "Istoriia fabrik i zavodov" series.

71. Gamel', *Opisanie*, 202.

72. Ibid., 199, 202. In 1937 V. E. Markevich repeated the assertion that Russia had standardized production with interchangeable parts of even its muzzleloaders; that is, earlier than in other European countries (*Ruchnoe ognestrel'noe oruzhie* [Leningrad, 1937], 184).

73. Gamel', *Opisanie*, xx; Solov'ev, *V pamiat'*, 30. Similar descriptions can be found, although less frequently, in the Soviet histories of Sestroretsk and Izhevsk. See *Sestroretskii instrumental'nyi zavod*, 28; and Gorbov, *Izhevskie oruzheiniki*, 27.

74. Q. 1371, 1372, 1375, 1376, *Report from the Select Committee on Small Arms*, 109. On Nasmyth, see Samuel Smiles, ed., *James Nasmyth: Engineer* (New York, 1883). Prosser's papers are at the Science Library in Lodon. Prosser began, but apparently never completed, an English translation of the study of Tula by Gamel'.

75. Q. 1422, 1434, *Report from the Select Committee*, 111–12.

76. Q. 2646, *Report from the Select Committee*, 172–73.

77. Q. 1422, Q. 1514, 1515, 1516, Q. 2790, *Report from the Select Committee*, 111, 116, 179.

78. Gamel', *Opisanie*, 117, 146.
79. Ibid., xviii.
80. Ibid., xii; Britkin, *The Craftsmen of Tula*, 38; *The Industries of Russia*, 150.
81. Aleksandrov, *Izhevskii oruzheinyi zavod*, 57–60.
82. Orfeev, "Istoriia Sestroretskogo oruzheinogo zavoda," 27–29; *Sestroretskii instrumental'nyi zavod*, 29, 35.
83. Gamel', *Opisanie*, xviii–xix. Emphasis added.
84. See the account by Gamel' of a demonstration to Nicholas I in *Opisanie*, xviii.
85. V. Chebyshev, "Po voprosu o novom administrativnom i tekhnicheskom ustroistve Tul'skogo oruzheinogo zavoda," *V.S.* 12 (1869): 249, recounts the tricks. Graf, "Oruzheinye zavody," 394, doubted their utility.
86. Britkin, *The Tula Craftsmen*, 39.
87. *Kuznitsa oruzhiia*, 66.
88. Orfeev, "Istoriia Sestroretskogo oruzheinogo zavoda," 18. Emphasis added.
89. Zagorskii, *A History of Metal Cutting Machines*, 237.
90. Britkin, *The Tula Craftsmen*, 31. The account is also in Afremov, *Istoricheskoe opisanie*, 211.
91. *Sestroretskii instrumental'nyi zavod*, 25–27; Britkin, *The Tula Craftsmen*, 28.
92. Zagorskii, *A History of Metal Cutting Machines*, 237.
93. Ibid., 238.
94. Ibid. It might also be added that weapons manufactured at the end of the eighteenth century greatly resembled those manufactured at the beginning of the century.
95. Ashurkov, "K istorii promyshlennogo perevorota," 53.
96. *Report from the Select Committee*, 350. The testimony is that of D. W. Witton, a gunmaker.
97. Q. 2647, 2650, 2697, 2698, *Report from the Select Committee*, 173, 175.
98. Subbotkin, "Ob Izhevskom oruzheinom zavode," 159.
99. Glebov, "Koe-chto," 164, 174, 179.
100. Quoted in S. G. Strumilin, *Chernaia metallurgiia v Rossii i v SSSR: Tekhnickeskii progress za 300 let* (Moscow, 1935), 222. The mining engineer is I. Kotliarevskii, writing in *Gornyi zhurnal* in 1873.
101. Orfeev, "Istoriia Sestroretskogo oruzheinogo zavoda," 30–31. Although the overall comparison of Russian and Western practices was valid, Orfeev's account contains a somewhat idealized version of western gunmakers, whose adaption to mechanized work was far less smooth than Orfeev suggests.

Chapter 5. America and the Russian Rifle

1. Hart, "Armed Forces," 305; Delafield, *Report*, 6–7; Kennett and Anderson, *The Gun in America*, 73–74; Dupuy, *Encyclopedia of Military History*, 731.

2. "Vzgliad na sostoianie russkikh voisk," *V.S.* 1 (1858): 1.
3. Ibid., 6–7.
4. Quoted in Fedorov, *Vooruzhenie* (1904), 155–56. Miliutin repeated the argument in his 1862 report. See Zaionchkovskii, *Voennye reformy*, 144.
5. Quoted in Zaionchkovskii, *Voennye reformy*, 56–57, 138.
6. V. Chebyshev, "Obozrenie russkikh voennykh zhurnalov," *V.S.* 10 (1861): 479–80, 488.
7. *A. Zh.* 1 (1857), 63; "Prikaz po Voennomu vedomstvu, no. 94 (20 March 1869)," *O.S.* 2 (1869): 1–4; *Vsepoddanneishii doklad po Voennomu ministerstvu* 8 (27 January 1871): 103.
8. *Ocherk preobrazovaniia v artillerii v period upravleniia generala-adiutanta Barantsova, 1863–1877 gg.* (St. Petersburg, 1877), 279–80; "Otchet za trekhletnee upravlenie artilleriei Ego Imperatorskim Vysochestvom General-Fel'dtseikhmeisterom," *A. Zh.* 1–6 (1860): 5: 297–98; Vorob'ev, *Novoe oruzhie*, 40; Zaionchkovskii, *Voennye reformy*, 136.
9. *A. Zh.* 1 (1857): 63.
10. *Ocherk preobrazovaniia*, 279–80; "Otchet za trekhletnee upravlenie," 5: 297–98; Vorob'ev, *Novoe oruzhie*, 40.
11. From the minutes of the Oruzheinyi otdel, 25–27 (1856), quoted in Fedorov, *Vooruzhenie russkoi armii XIX v.*, 131.
12. From the Miliutin papers in the Manuscript Division of Lenin Library, quoted by P. A. Zaionchkovskii, *Voennye reformy*, 139. Miliutin's assertion that inventions were "suddenly forgotten" is somewhat disingenuous: tests continued even as the system tested became outmoded (see A. A. Smychnikov and Val. V. Mavrodin, "K voprosu o perevooruzhenii russkoi armii v seredine XIX veka," *Problemy istorii feodal'noi Rossi* [Leningrad, 1971], 248.)

By the mid-1860s, the Russian army also began to show greater interest in rapidity of fire and, even before 1860, had begun tests on breechloaders. One of the earliest breechloading systems seriously considered in Russia was that of the American Green. In 1859 Green offered to the Ministry of War a double-barreled breechloading cavalry carbine that used a paper cartridge and a cylindrical bullet. Early tests at the School of Sharpshooters at Tsarskoe Selo, outside St. Petersburg, were very favorable, especially for accuracy. Russia ordered 3,000 Green carbines for the dragoons. However, later tests conducted by the Kuban Cossacks were not as favorable, and the Green cavalry carbine was never officially adopted. See "Po voprosu o vooruzhenii nashei armii skorostrel'nymi ruzh'iami," *V.S.* 11 (1866): 35; "Perechen' zaniatii Oruzheinoi komissii," *A. Zh.* 2 (1857): 134; 11 (1857) 179; 3 (1862): 149–50; 4 (1862): 31–42; "Otchet za trekhletnee upravlenie," 310–11; A. Svistunov, "Publichnye lektsii, chitannye pri gvardeiskoi artillerii v 1862 g.," *O.S.* 1 (1862): 103; Capt. Kovalevskii, "Ob ispytanii ruzhei dvukh-pul'noi sistemy Grina," *O.S.* 1 (1865): 17; Capt. Paramanov, "Otchet o sravnitel'nom ispytanii ruzhei dvukhpul'noi sistemy Grina, s 6-lineynymi kazach'imi vintovkami, v uchebnoi sotne Kubanskogo kazach'ego voiska," *O.S.* 4 (1867): 68.

13. *Ocherk preobrazovanii,* 286–87.
14. Terry carbines were already familiar to the small arms commission. See "Perechen' zaniatii oruzheinoi komissii," *A. Zh.* 7 (1861): 309–10, 9 (1861): 397–98; "Pravitel'stvennoe rasporiazhenie," *O.S.* 4 (1866): 48–75; Orfeev, "Istoriia Sestroretskogo oruzheinogo zavoda," 52. The advantages of the Terry-Norman were its sturdy lock mechanism and the cover that protected the breechblock from dirt and rain (Fedorov, *Vooruzhenie russkoi armii XIX v.,* 159.) On the shortcomings, see P. Bil'derling and V. Buniakovskii, "Russkaia igol'chataia vintovka," *O.S.* 1 (1868): 3; Pototskii and Shkliarevich, *Sovremennoe ruchnoe oruzhie,* 335–85; Ashurkov, *Tul'skii muzei oruzhiia: Putevoditel'* (Tula, 1972), 33, 35.
15. "Zhurnal oruzheinoi komissii Artilleriiskogo komiteta za No. 75 o proizvodstve opytov nad oruzhiem zariazhaiushimsia s kazennoi chasti i deistvuiushimi gotovymi metallicheskimi patronami," *O.S.* 4 (1864): 19–28; "Pravitel'stvennoe rasporiazhenie v prikaze po artillerii za No. 45," *O.S.* 1 (1868): 13; *A. Zh.* 2 (1868): 371; Bil'derling and Buniakovskii, "Russkaia igol'chataia vintovka," 4; Pototskii and Shkliarevich, *Sovremennoe ruchnoe oruzhie,* 335–85; V. Eksten, *Opisanie sistem skorostrel'nogo oruzhiia* (Moscow, 1870), 25, 41; Fedorov, *Vooruzhenie russkoi armii XIX v.,* 159; Greene, *The Russian Army,* 1: 52; Fedorov, *Vooruzhenie* (1904), 89–90; *Ocherk preobrazovaniia,* 289–90.
16. "Pravitel'stvennoe rasporiazhenie No. 14," *O.S.* 3 (1869): 11; "Prikaz po Voennomu vedomstvu No. 270," *O.S.* 4 (1869): 7; "Perechen' zaniatii ispolnitel'noi komissii po perevooruzheniiu armii za 1869 g.," *O.S.* 1 (1870): 38–85, 2 (1870): 54–108; Greene, *The Russian Army,* 1: 52, 54; E. B. Hamley, "The Armies of Russia and Austria," *Nineteenth Century* 3 (May 1878): 852; Mavrodin and Mavrodin, *Iz istorii,* 65–67; Beskrovnyi, *Russkaia armiia i flot,* 303; Ashurkov, "K istorii promyshlennogo perevorota Rossii," 116–17; *Small Arms of the World,* 10th ed., 60. Among other things, the trajectory was high and the extractor to eject the empty cartridges caused great difficulties.
17. "Amerikanskaia artilleriia," *Inzhenernyi zhurnal* 4 (1864): 9.
18. P. Lukin, "Novye obraztsy oruzhiia, zariazhaiushegosia s kazni," *O.S.* 1 (1865): 6–8; Eksten, *Opisanie sistem,* 105–14; *Winchester Repeating Arms Company* (New Haven, 1869), 54–55; V. P. V. Vorob'ev, "Kakoe oruzhie nam nuzhno," *O.S.* 4 (1867): 116.
19. Eksten, *Opisanie sistem,* 92; Kennett and Anderson, *The Gun in America,* 73–74.
20. Eksten, *Opisanie sistem,* 92.
21. "Gorlov, A. P.," *Voennaia entsiklopediia,* 18 vols. (St. Petersburg, 1911-1915), 8: 403; *A. Zh.* 1 (1859): 48–59; V. Chebyshev, "Opisanie ustroistva metallicheskikh patronov, priniatykh dlia voennogo oruzhiia i posledovatel'nogo ikh usovershenstvovoniia," *O.S.* 1 (1872): 29; Russian Legation to Department of State, 31 October 1868, Washington, "Notes from the Russian Legation in the U.S. to the Department of State, 1809–1906," roll 4, Microcopy no. 39, NA.

22. Chebyshev, "Opisanie ustroistva," 29; Aleksandr Gorlov, "Ob upotrebliaemykh v armii Soedinennykh Amerikanskikh Shtatov ruzh'iakh, zariazhaiushikhsia s kazennoi chasti, i k nim metallicheskikh patronov," *O.S.* 3 (1866): 13–16; Fedorov, *Vooruzhenie russkoi armii XIX v.*, 162.

23. Pototskii and Shkliarevich, *Sovremennoe ruchnoe oruzhie*, 363; Fedorov, *Vooruzhenie russkoi armii XIX v.*, 159, 163–64; Zaionchkovskii, *Voennye reformy*, 170; A. Pozdnev, *Tvortsy otechestvennogo oruzhiia* (Moscow, 1955), 123; Mavrodin and Mavrodin, *K istorii*, 73. Stoeckl to Swift, 21 November 1866, Washington, Box 9, Folder S, CPFAM Records, RG 103, CSL. An account of questionable reliability states that when Gorlov and Gunius arrived in the United States in January 1868 on their "secret" mission, "the two officers got in touch with the Russian embassy in Washington, which referred them to the Colt firm, whose president, Samuel Colt, welcomed them most cordially in Hartford, Connecticut." Admittedly, Colt had become a legendary figure, but his death in 1862 most likely precluded such a welcome (Tsaraidze, *Czars and Presidents*, 32–33).

24. Eksten, *Opisanie sistem*, 155.

25. Gorlov, "Ob upotrebliaemykh ruzh'iakh," 17, 24–25; Eksten, *Opisanie sistem*, 153.

26. Eksten, *Opisanie sistem*, 152; Gorlov, "Ob upotrebliaemykh ruzh'iakh," 13–16.

27. For the first information about Berdan appearing in Russia, see "Ruzh'e Berdana," *O.S.* 3 (1867): 35–39; and "Berdan," *Voennaia entisiklopediia*, 4: 480.

28. Eksten, *Opisanie sistem*, 154–55.

29. From unpublished papers in the archives of the Ministry of War, cited by Zaionchkovskii, *Voennye reformy*, 170. See also "Perechen' zaniatii oruzheinoi komissii," *O.S.* 4 (1867): 27; "Igol'chatye ruzh'ia v zapadnoi Evrope i u nas," *Russkii invalid*, 254 (6/18 October 1866): 3; N. Litvinov, "Nasha peredelochnaia shestilineinaia vintovka po sisteme Krynka i patron Berdana," *Russkii invalid* 132 (6/18 November 1869): 3; Mavrodin and Mavrodin, *Iz istorii*, 73; Fedorov, *Evoliutsiia*, 1: 117–18.

30. *Ocherk preobrazovanii*, 295–95; Gnatkovskii, *Istoriia*, 115; Pozdnev, *Tvortsy* 123; V. Buniakovskii, "Neskol'ko slov o svoistvakh russkoi 4.2-lineinoi vintovki," *O.S.* 4 (1869): 1–25.

31. "Nastavlenie dlia osmotra russkoi malokalibernoi vintovki, zariazhaiushcheisia s kazennoi chasti," *O.S.* 4 (1868), 53; Eksten, *Opisanie sistem*, 160. Schuyler to Department of State, no. 59, 10 March 1869, Moscow, "Despatches from U.S. Consuls in Moscow, 1857–1906," roll 1, Microcopy no. 456, NA. According to a Soviet historian with access to unpublished records, the licensing agreement and Gorlov's report to the Main Artillery Administration are located in the Military-Historical Archives. See M. K. Portnov, "K istorii priniatiia na vooruzhenie russkoi armii 4.2-lineinoi vintovki obraztsa 1868 g." in *Ezhegodnik Gosudarstvennogo istoricheskogo muzeia 1961 g.* (Moscow, 1962), 69-70.

32. "Doklad, podannyi Glavnym Artilleriiskim Upravleniem Voennomu Ministru 24 noiabria 1866," cited in Fedorov, *Vooruzhenie russkoi armii XIX v.,* 163. See also V. Chebyshev, "Opisanie ustroistva, " 30, 63; Eksten, *Opisanie sistem,* 150. More detailed technical description in English of the Berdan rifle can be found in *Small Arms of the World,* 54, and in Tarassuk and Blair, *Complete Encyclopedia,* 227. Ironically, Russia adopted the Berdan small-caliber infantry rifle five years before the U.S. Army adopted the Springfield small-caliber rifle (Mavrodin and Mavrodin, *Iz istorii,* 73).

33. William Jarvis to CPFAM, 3 April and 14 September 1868, Hartford, CPFAM Records, RG 103, CSL.

34. V. Buniakovskii, "Ustroistvo Kol'tovskogo zavoda," *O.S.* 4 (1869), 53–59, 64.

35. Eksten, *Opisanie sistem,* 155; Buniakovskii, "Ustroistvo," 53, 59.

36. "Novoe ruzh'e i patron generala Berdana," *O.S.* 3 (1869): 32–40; N. Litvinov, "O sravnitel'nom ispytanii proizvedennom v uchebnom pekhotnom batalione, ruzhei sistem Berdana (novoi) i Verdera (Bavarskoi)," *O.S.* 3 (1869): 56–67, 4 (1869): 80–122; "Opisanie russkoi malokalibernoi vintovki so skol'ziashchim zatvorom vtoroi sistemy Berdana, ee razborka, sborka, chistka, sberezhenie i pr.," *O.S.* 2 (1872): 88–111. Dispatch of Wellesley, 26 October 1869 and 22 January 1873, St. Petersburg, Foreign Office Records, FO 65/786 and FO 519/274 PRO; Berdan to Franklin, 6 November 1870, St. Petersburg, Box 4, "Franklin Folder," CPFAM Records, RG 103, CSL; Pototskii and Shkliarevich, *Sovremennoe ruchnoe oruzhie,* 1: 371–72; 2: 250; *Ocherk preobrazovaniia,* 299–300; Greene, *The Russian Army,* 53; Ashurkov, *Tul'skii muzei oruzhiia,* 36–37; Zaionchkovskii, *Voennye reformy,* 177.

37. Berdan to Franklin, 6 November 1870, St. Petersburg, Box 4, "Franklin Folder," CPFAM Records, RG 103, CSL. According to Fedorov, the Berdan 2 sliding bolt action rifle had the following advantages over the Berdan 1 trap door action and other sliding bolt actions tested in Russia at the time: easier to manufacture, easier to assemble and to disassemble, easier to clean, easier to load, easier to extract the cartridge case, a better sight, greater safety, and greater rate of fire (*Evoliutsiia,* 1: 122–23.)

38. Greene, *The Russian Army,* 53; Fedorov, *Vooruzhenie russkoi armii XIX v.,* 245; Fries, "A Comparative Study," 239–41.

39. Berdan to Franklin, 28 September 1870, St. Petersburg, Box 4, "Franklin Folder," CPFAM Records, RG 103, CSL.

40. Franklin to CPFAM, 3 December 1869, Hartford, Box 14, Folder "1858–1869," CPFAM Records, RG 103, CSL; Von Oppen to CPFAM, 18 May 1872, London, Box 14, CPFAM Records, RG 103, CSL.

41. V. Shkliarevich, "Sovremennoe sostoianie voprosa o vooruzhenii pekhoty," *Russkii invalid* 25 (22 February 1869): 5. See also Pototskii and Shkliarevich, *Sovremennoe ruchnoe oruzhie,* 363.

42. Litvinov, "Opisanie skorostrel'noi vintovki Berdana," *O.S.* 3 (1869): 31.

43. Quoted in Mavrodin and Mavrodin, *Iz istorii,* 77.

44. Tsarsaidze, *Czars and Presidents*, 30.
45. Norton, *American Inventions and Improvements*, 57, 96–105. See also John Latham, "Progress of Small Breech-loading Arms," *Journal of the Royal United Service Institute 9*, 83 (1875): 643.
46. Needless to say, Soviet historians writing in the 1950s were not impressed by the foreign mystique and claimed that, despite the authorship of Gorlov and Gunius, the Russian government stubbornly insisted Berdan was the inventor and paid him 50,000 rubles. "The small calibre rifle invented by Russian officers was sold twice to the tsarist government by American opportunists." This story transpired because the government was allegedly proud that the Russian army had rifles bearing non-Russian names. See Y. I. Sirota, "Perevooruzhenie russkoi armii vo II polovine XIX v." kand. diss. (Leningrad, 1950), 23; and Pozdnev, *Tvortsy*, 124. To their credit, Zaionchkovskii, *Voennyi reformy*, 176), and V. V. Mavrodin, "K voprosu o perevooruzhenii russkoi armii v seredine XIX v.," in *Problemy istorii feodal'noi Rossii* (Leningrad, 1971), 90–93, relate the story in a more dispassionate way.
47. In 1869 a committee of the U.S. Army tested several models of firearms, including the Colt-made Berdan 1. Records of the committee's tests from the Artillery Department indicated that the Russian Berdan successfully passed all tests yet placed behind the Springfield, Sharps, and Remington models. Although the Remington displayed defects not published in the committee's record, its prior adoption by the Navy suggested to Gorlov that the tests in St. Louis were organized to justify a decision already made in favor of the Remington. He blamed the surprisingly low placement of the Berdan on the fact that, because Berdan had sued the U.S. Army for patent violations in 1866, many "American military and government figures want to have nothing to do with Berdan, and any rifle with his name attached to it would not fare well." Finally, with a degree of national pride, Gorlov claimed that it would have been embarrassing to the U.S. Army to give the highest marks to a rifle designed by Russian officers. See "Memoriia artilleriiskogo upravleniia SShA ob ispytanii skorostrel'nykh ruzhei," *O.S.* 4 (1871): 11–12, 16. For further tests on the Berdan in Russia, see "Ispytanii ruzhei Berdana–starogo i novogo obraztsa–i Verdera," *O.S.* 2–4 (1870); "Prodolzhenie opytov nad malokalibernym ruzh'iam vtorogo obraztsa," *O.S.* 3 (1871): 27–56; and "Opyty nad malokalibernymi ruzh'iami Berdana, vtorogo obraztsa," *O.S.* 1 (1872): 26–48.
48. V. Buniakovskii, "O metallicheskikh patronakh k russkoi malokalibernym vintovkam," *O.S.* 3 (1870): 53–54.
49. Norton, *American Inventions and Improvements*, 309–10.
50. Gorlov to Franklin, 29 July 1869, Hartford, Box 4, Folder G, CPFAM Records, RG 103, CSL; V. Buniakovskii, "O metallicheskikh patronakh k russkoi malokalibernym vintovkam," *O.S.* 3 (1870): 53–54.
51. Hatch, *Remington Arms*, 115. According to Hatch, writing in 1956, a Russian was sent to spy on Gorlov by getting a job at UMC. To Hatch, this accounted for the fact that, although Gorlov was a "polished gentleman,"

"Russians who served their government then were almost as fear-ridden as Communists are today," (*Remington Arms*, p. 115).

52. The Grand Duke also observed torpedo tests, feasted on a banquet catered by Delmonicos in New York, and received a silver-inlaid snuff box depicting the parts of the Russian metallic cartridge. "Poseshchenie velikim kniazem Alekseem Aleksandrovichem Bridgzhportovskogo patronnogo zavoda," *A. Zh.* 2 (February 1872): 237–40; Hatch, *Remington Arms*, 116–17.

53. A. Fon-der-Khoven, "Materialy dlia ocherka razvitiia ruzheinoi fabrikatsii v SShA," *O.S.* 4 (1877): 29. The story is repeated in Norton, *American Inventions and Improvements*, 300; N. Litvinov, "Nasha peredelochnaia shestilineinaia vintovka," 3.

54. Gorlov, "Ob upotrebliaemykh ruzh'iakh," 32–33.

55. A description of the Civil War in a Russian city in a recent Soviet novel. Iurii Trifonov, "Starik," *Druzhba narodov* 3 (1978): 38.

56. C. E. H. Vincent, "The Russian Army," *Journal of the Royal United Service Institution*, 16, 67 (1872): 302–3.

57. "Perechen' zaniatii oruzheinoi komissii," *O.S.* 2 (1868): 11. The published sources, reticent enough regarding military affairs, are virtually silent about adoption of weapons by the police and gendarmes.

58. Fiedler to Smith and Wesson, 8/21 May 1871, St. Petersburg, Russian contracts and correspondence, Smith and Wesson Records, Private Collection of Roy Jinks. The letter includes an order for 160 pistols. Fiedler confesses in another letter, "The new No. 3 Revolver is very pretty indeed, but I have not yet found success; my customers think the price too high and the size too heavy" (30 March/10 April 1871). Lukin, "Novye obraztsy oruzhiia, zariazhaiushchegosia s kazni," *O.S.* 1 (1865): 8–11. "Perechen' oruzheinoi komissii," *O.S.* 3 (1865): 38; Artilleriiskii istoricheskii muzei, *Putevoditel' po istoricheskomu arkhivu muzeia* (Leningrad, 1957), 157. See also L. P., "Revol'ver, kak oruzhie dlia kavalerii," *V.S.* 9 (1866): 25–52. Berdan to Franklin, 17 October, St. Petersburg, Box 14, CPFAM Records, RG 103, CSL. D. Miliutin, "Vsepoddanneyshii doklad po Voennomu ministerstvu," for the years 1870 and 1871, cited in D. Skalon, comp., *Stoletie Voennogo ministerstva*, 13 vols. (St. Petersburg, 1902–1914), Appendix to vol. 1: 105, 179; "Vooruzhenie nashei kavalerii," *O.S.* 2 (1871): 146.

59. Roy Jinks in conversation with the author, Springfield, Mass., 28 March 1979. In particular, the contract signed on 15 December 1873 states, concerning interchangeability, that the work must be "equal in this respect to the work upon firearms made at any establishment in the world." The company correspondence includes frequent additions to contracts for a change of detail. For Russian sources, see also "Memoriia artilleriiskogo upravleniia SShA ob ispytanii skorostrel'nykh ruzhei," *O.S.* 3 (1871): 17; "Pravitel'stvennoe rasporiazhenie v tsirkuliarakh Glavnogo Shtaba, nos. 4, 51, 61," *O.S.* 2 (1874): 8–10; no. 1, *O.S.* 1 (1875): 1; and no. 116, *O.S.* 2 (1878): 8–9; "Perechen' zaniatii Oruzheinogo otdela Artilleriiskogo komiteta, nos. 51, 63, 73," *O.S.* 4 (1876): 29, 41, 48; *Podrobnyi alfavitnyi*

ukazatel' prikazov po Voennomu vedomstvu s tsirkuliarami Glavnogo Shtaba (St. Petersburg, 1876). See also Greene, *The Russian Army*, 73–75; Roy G. Jinks, *Smith and Wesson, 1857–1945* (New York, 1975), 67, 121–22; Carl Reinhold Hellstrom, "Smith and Wesson: One Hundred Years of Gunmaking! 1852–1952," quoted by Martin Rywell, *Samuel Colt: A Man and an Epoch* (Harriman, Tenn., 1952), 75.

60. Norton, *American Inventions and Improvements,* 200.

61. Gorloff to Wesson, 15 February 1872, Springfield Russian Contracts and Correspondence, Smith and Wesson Records, Private Collection of Roy Jinks.

62. Conversation with the author, Springfield, Mass., 28 March 1979. See also Jinks, *Smith and Wesson,* 68.

63. Gorlov to Franklin, 21 November 1881, London, Box 14, Von Oppen Folder 1881, CPFAM Records, RG 103, CSL. Jinks's calculation up to 1877 is 131,138 (Jinks, *Smith and Wesson,* 103); Pototskii, *Sovremennoe ruchnoe oruzhie* (1904) 34. Filatov and N. Iurlov, "O trekhlineinom revol'vere obraztsa 1895 g. i opisanie revolvera," *O.S.* 1 (1896): 1–46. Occasional orders for Smith and Wesson may have continued until the beginning of the twentieth century. See Artilleriiskii istoricheskii muzei, *Putevoditel',* 119. Over a period of several years it was issued to infantry and artillery officers; sergeant-majors, drummers, and buglers; cuirassiers and front-rank lancers of cavalry; and, by 1881, naval officers. Although it was also issued to gendarme officers and to convoy and prison guards, little material is available on the use of the revolver by the police. See "Pravitel'stvennoe rasporiazhenie v prikazakh po Voennomu vedomstvu, no. 256," *O.S.* 4 (1876): 3; and no. 58, *O.S.* 3 (1880): 2; "Prikazy upravliashchego Morskim ministerstvom no. 40," *Morskoi sbornik* 190, 5 (May 1882): 29; K. Patin, comp., *Spravochnik: Polnyi i podrobnyi ukazatel' vsekh deistvuiushchikh prikazov po voennomu vedomstvu* (Tambov, 1904), 2: 60, 61, 86, 451–52; *Sistematicheskii sbornik prikazov po voennomu vedomstvu i tsirkuliarov Glavnogo shtaba* (St. Petersburg, 1886), 245–46; "Izvlechenie iz otcheta Artilleriiskogo otdeleniia Morskogo tekhnicheskogo komiteta: O vvedenii vo flot revol'verov Smita i Vessona," *O.S.* 4 (1882): 21–24; "Perechen' zaniatii Oruzheinogo otdela Artilleriiskogo komiteta," *O.S.* 2 (1880): 81–82.

64. By the mid-1880s, all ranks of the cavalry were armed with the revolver and "like the American, but contrary to our own and other continental armies, they are trained to depend to a considerable extent on it." One English observer even claimed that the English cavalry might not be a match for the Russians (H. E. C. Kitchener, "Revolvers and their Use," *Journal of the United Service Institute* 30, 136 [1886]: 991). Reviewing the Vienna Exposition of 1873, Aleksandr I. Fon-der-Khoven, a specialist in small arms, observed that the Smith and Wesson small pocket revolvers were eagerly purchased on the spot and everywhere one heard praise both for the revolver system and for its beauty and flawless execution. See "Zametki o venskoi vsemirnoi vystavke 1873 g.," *O.S.* 1 (1874): 42; *Otchet*

po venskoi vsemirnoi vystavke 1873 g. v voenno-tekhnicheskom otnoshenii (St. Petersburg, 1879), 285; Moskovskaia politekhnicheskaia vystavka, *Ukazatel' kollektsii artilleriiskogo otdela Moskovskoi politekhnicheskoi vystavki 1872 g.* (Moscow, 1872), 85. For later criticism, see V. Goncharov, "Revol'ver Mervina po sravneniiu s revol'verom Smita i Vessona," *O.S.* 4 (1899): 47–48; and A. Popov, "Neobkhodimost' zameny revol'verov Smita i Vessona," *O.S.* 4 (1891): 2–3.

65. Gatling Gun Company to Gorlov, 12 and 25 July 1871, Hartford, Box 36, Folder "Gatling–Contract and Letters with Russian Government, 1869–72," CPFAM Records, RG 103, CSL; Zaionchkovskii, *Voennye reformy*, 161; Fedorov, *Evoliutsiia*, 135–40, 180–90; Vincent, "The Russian Army," 306.

66. Vincent, "The Russian Army," 306.

67. "Pravitel'stvennye rasporiazheniia v prikazakh po artillerii, no. 40," *O.S.* 3 (1872): 23; "Igol'chatye ruzh'ia," 3; Andrew Buchanan to Granville, no. 91, 17 May 1871, London, Foreign Office Records, FO 65/821, PRO.

68. Quoted in William C. Askew, "Russian Military Strength on the Eve of the Franco-Russian War," *Slavonic and East European Review* 30 (1951):200.

69. Miliutin, "Doklad" (1862), cited in Skalon, *Stoletie*, 1: 135–36; Wilhelm Ploennis, *Novye issledovaniia nad nareznym pekhotnym ognestrel'nym oruzhiem* (St. Petersburg, 1863), 68–69.

70. *Ocherk preobrazovaniia*, 315; Chebyshev, "Peremena obraztsa dlia peredelki nashikh ruzhei," *O.S.* 2 (1869): 1–2; V. Shklarevich, "Sanktpeterburgskaia masterskaia metallicheskikh patronov," *Russkii invalid* 23 (22 February 1869): 3–4; Eksten, *Opisanie sistem*, 152.

71. N. Litvinov, "Opisanie skorostrel'noi vintovki Berdana," *O.S.* 3 (1869): 109–10. See also the documents cited by Fedorov, *Vooruzhenie russkoi armii v XIX v.*

72. Cited in Fedorov, *Vooruzhenie russkoi armii v XIX v.*, 237. Emphasis in the original.

73. Menning, *Bayonets Before Bullets*, 5.

74. Ibid., 34–35; Strokov, *Istoriia voennogo iskusstva*, 1: 611–12.

75. Voenno-istoricheskaia komissiia Glavnogo shtaba, *Opisanie russko-turetskoi voiny 1877–1878 gg. na Balkanskom poluostrove* (St. Petersburg, 1901), 1: 123. The best recent account of Russian military thought is Meshcheriakov, *Russkaia voennaia mysl'*.

76. M. I. Dragomirov, *Sbornik original'nykh i peredovykh statei, 1858–1880*, 2 vols. (St. Petersburg, 1881), 1: 302; M. I. Dragomirov, "Lektsii taktiki, chitannye v uchebnom pekhotnom batal'one," *O.S.* 2 (1864): 17.

77. Dragomirov, *Sbornik original'nykh i peredovykh statei*, 1: 295.

78. M. I. Dragomirov, "Vliianie rasprostraneniia nareznogo oruzhiia na vospitanie i taktiku voisk," *O.S.* 1 (1861): 53, 94. See also Menning, *Bayonets Before Bullets*, 31.

79. Menning, *Bayonets Before Bullets*, 31; Strokov, *Istoriia voennogo iskusstva*, 1: 618–21; Dragomirov, "Vliianie," 85, 298. See also Meshcheriakov, *Russkaia voennaia mysl'*, 198.

80. Dragomirov, "Vliianie," 74.

81. Strokov, *Istoriia voennogo iskusstva*, 618–21; Dragomirov, "Vliianie," 492–93. See also his "Vintovki, skorostrel'naia i obyknovennaia," *O.S.* 2 (1869): 32; and his "O veroiatnykh peremenakh v taktike vsledstvie rasprostraneniia dal'no- i skorostrel'nogo oruzhiia," *V.S.* 11 (1867), 16.

82. Pintner, "Russian Military Thought," 368; Von Wahlde, "Military Thought," 137–38; C. E. Mansfield to the Foreign Secretary, 25 July 1866, St. Petersburg, Foreign Office Records, FO 65/708, PRO; Meshcheriakov, *Russkaia voennaia mysl'*, 198; *Ocherk preobrazovaniia*, 284.

83. Von Wahlde, "Military Thought," 154; Menning, *Bayonets Before Bullets*, 29. In fairness to Russian officers, as Showalter demonstrates, this thinking was widespread in continental armies (*Railroads and Rifles*, 103, 109–12, 123).

84. Menning, *Bayonets Before Bullets*, 32–34; Meshcheriakov, *Russkaia voennaia mysl'*, 204.

85. Graham, "The Russian Army," 233–34.

86. No proponent of the new weapons, Dragomirov realized that the concept of military education would need to change. New subjects would enter standard military training — marksmanship, fencing, gymnastics, obstacle courses. See Dragomirov, "Vliianie," 58; Fedorov, *Vooruzhenie* (1904), 156; "Otchet za trekhletnee upravlenie artilleriei," *A. Zh.* 5 (1860): 336 1 (1855): 43; A. Verkhovskii, *Ocherk po istorii voennogo iskusstva v Rossii XVII–XIX vv* (Moscow, 1921), 217. See also Pintner, "The Burden of Defense," and Menning, *Bayonets Before Bullets*, 34–35.

87. Dragomirov, "Vliianie," 37. In Europe, too, breechloaders were not trusted in the hands of the untrained soldier (Showalter, *Railroads and Rifles*, 85, 93, 103, 124). See also Vorob'ev, "Zametki o revol'verakh," *O.S.1* (1869): 5, 19; V. F. Argamakov, "Vospominaniia o voine 1877–1878 gg.," *Zhurnal Imperatorskogo Russkogo voenno-istoricheskogo obshchestva* 2–7 (1911)7: 181.

88. Fedorov, *Vooruzhenie* (1904), 5. In fairness to the much maligned peasant soldier, one foreign observer believed that the greater amount of technical training would prove no obstacle since the younger recruits were already better disciplined and eager to learn. "It is only necessary to see, as one can frequently see in Moscow, the recruits who are brought in from the country and the same men after six months to perceive the great benefit which they derive from the instruction they receive in garrison schools, in the habits of order and neatness and in book knowledge. Reports which are frequently repeated in foreign journals of their utter stupidity and inability to learn the use of breech-loading guns are utterly false; as far as my experience goes the Russian peasantry are far brighter and quicker to learn than the same classes in Germany or Ireland" (Schuyler, no. 59, 10 March 1869, Moscow, "Despatches from U.S. Consuls in Moscow, 1857–1906," roll 1, Microcopy no. 456, NA).

89. *Istoriia armii i flota*, 11: 43.

90. Konstantinov, "Posledovatel'nye usovershenstvovaniia," 52–53. See also the textbook *Kurs taktiki* (St. Petersburg, 1859), 175–76.

91. N. Egershtrom, "Svedeniia, otnosiashchiesia do vvedeniia v russkuiu armiiu ruchnogo oruzhiia umen'shennogo kalibra," *O.S.* 1 (1861): 49; A. Vel'iaminov-Zernov, "Teoreticheskii kurs o ruchnom ognestrel'nom oruzhii, prepodavaemyi pri uchebnom pekhotnom batalione," *O.S.* 1 (1864): 30–31. See also M. Terent'ev, "Zametki o vooruzhenii kavalerii," *V.S.* 5 (1858): 70.

92. M. Terent'ev, "Kavaleriiskie voprosy," *V.S.* 8 (1865): 285–88; Konstantinov, "Posledovatel'nye usovershenstvovaniia," 52–53. See also the textbook *Kurs taktiki* (1859), 175–76. Likewise, V. P. Vorob'ev argued that the revolver was only a defensive weapon ("Revol'ver, kak oruzhie stroevogo ofitsera," *O.S.* 1 (1866): 3). See also Meshcheriakov, *Russkaia voennaia mysl'*, 211–13; Graham (1883), "The Russian Army, 247.

93. Gorlov, "Ob upotrebliaemykh ruzh'iakh," 8–10. Eksten, *Opisanie sistem*, 5, also borrows this argument. See also Fedorov, *Vooruzhenie* (1904), 121.

94. Wellesley to Foreign Office, no. 18, 12 December 1871, St. Petersburg, Foreign Office Records, FO 519/274, PRO. See also Grierson, *The Armed Strength of Russia* (London, 1873), 112–13; Zaionchkovskii, *Voennye reformy*, 139. French sources calculated that Russia had 870,000 breechloaders. See *Revue d'Artillerie* 2, 4 (April–September 1874): 92–93.

95. "Prikaz po Voennomu vedomstvu, no. 287 and no. 334," *O.S.* 4 (1869): 7–9; "Prikaz po Voennomu vedomstvu," no. 94 and no. 261, in Voennoe ministerstvo, *Peremeny v obmundirovanii i vooruzhenii voisk, 1855–1881*, 25 vols. (St. Petersburg, 1857–1181), 90 (1871): 59–60, and 91 (1871): 70–72; Pototskii and Shkliarevich, *Sovremennoe ruchnoe oruzhie*, 385; J. M. Grierson, *The Armed Strength of Russia* (London, 1873), 113–14; or 2d ed. (London, 1883), 105–8; C. E. H. Vincent, "Russian Army," *Journal of the Royal United Service Institute* 16, 67 (1872): 298; Graham, "The Russian Army," 232; K. D. Bubenkov, "Perevooruzhenie russkoi armii v 1860–1870 gg.," kand. diss. (Leningrad, 1952), 21; Zaionchkovskii, *Voennye reformy*, 178; *Istoriia armii i flota*, 11: 40–41; 13: 26–29. As late as 1877 the Caucasus, Orenburg, Siberian, and Turkestan military districts had the 1867 model Carl rifle with paper cartridges. The Moscow, Kiev, Kharkov, Odessa, Kazan, Finland, Vilno, and part of the Caucasus military districts had the Krnka rifle. The British, understandably watchful over Russian armament, reported that, although the intention was eventually to issue 200,000 Berdans to the Cossacks, as of 1871 only 20,000 had been ordered and would be supplied. By 1876 Cossack units had received a small number of the Berdan smallcaliber rifles to replace the Krnka. Most divisions still had .60 caliber muzzleloaders and some still had smoothbore flintlocks. Although dragoons were supposed to have the second Berdan, in fact only the guard dragoon companies had the new weapon, while the rest had Krnkas. Distribution of Smith and Wesson revolvers proceeded more quickly, no doubt because distribution was primarily to officers (Voennoe ministerstvo, Voennaia istoricheskaia komissiia Glavnogo shtaba, *Opisanie russko–turetskoi voiny*, 1: 122–23, 126–27; Grierson, *Armed Strength*, 103–22; Col. Blane to Buchanan, no. 3, 23 January 1871, St. Petersburg, Foreign Office Records, FO

65/820, PRO); S. Mikheev, *Istoriia russkoi armii*, 5 vols. (Moscow, 1910–1911), 5: 14–15. See also Argamakov, "Vospominaniia," 4:77. Russia may have been better armed than Turkey in Central Asia. Although the Russian infantry generally had the obsolete Carl, the Turkish forces had flintlocks, percussion muskets, double-barreled shotguns, and even some matchlocks; many units had no more than sabers (Grishinskii, *Istoriia russkoi armii i flota*, 12: 113). According to Beyrau, Turkish field artillery, supplied by Krupp, proved superior to Russian field artillery (*Militär und Gesellschaft*, 318–20).

96. At the outbreak of the war, Turkey had 325,000 Snider converted breechloading infantry rifles and 334,000 Peabody-Martini breechloaders (Grishinskii, *Istoriia russkoi armii i flota*, 11:36). According to official unpublished records cited by Zaionchkovskii, Russia had 150,868 Carl needle rifles, 372,700 Krnka infantry rifles, 270,962 Berdan 1 and 2 infantry rifles, and 74,454 Berdan carbines and dragoon and Cossak rifles (*Voennye reformy*, 178). For more on Russian and Turkish arms see Grishinskii, *Istoriia russkoi armii i flota*, 11: 36, 40–41, 161; 13: 29–34; N. P. Pototskii, *Turetskie ruzh'ia vo vremia voiny 1877 g. Sravnenie turetskogo vooruzheniia s russkimi* (St. Petersburg, 1878), 10–11; Norton, *American Improvements*, 62; Eksten, *Opisanie sistem*, 94–104; A. Fon-der-Khoven, "Zametka o Providenskom oruzheinom zavode i o ruzh'iakh Pibodi-Martini, Turetskogo obraztsa," *O.S.* 4 (1877): 3, 65. See also Verkhovskii, *Ocherk po istorii voennogo iskusstva v Rossi XVII–XIX vv.* (Moscow, 1921), 220. Many foreigners concurred. See F. V. Greene, *Report on the Russian Army and its Campaigns in Turkey in 1878* (New York, 1879), 40. Pototskii argued that stories to the effect that the Turks were better armed were unfounded (*Turetskie ruzh'ia*, 120; "O znachenii bystroi strel'by, o noveishikh obraztsakh magazinnykh ruzhei i priborakh dlia uskoreniia ruzheinoi strel'by," *O.S.* 4 [1879]: 4). The editors of *Oruzheinyi Sbornik* noted that the praise for the Peabody-Martini emanated from the general, nonspecialized Russian press. To correct this allegedly erroneous impression, the editors published in 1869 a French brochure based largely on data from English tests to demonstrate that the Peabody-Martini, also referred to as the Henry-Martini, the English equivalent, was no better than the Berdan ("Zametki o ruzh'e Genri-Martini, predlozhennom k priniatiiu angliiskogo pravitel'stva osoboi komissiei," *O.S.* 1 [1878]: 115–42). On the early use of the magazine rifle, see S. A. Zybin, "Perevooruzhenie ruzh'iami umen'shennogo kalibra," *O.S.* 3 (1894): 43–44; *Opisanie*, 1:33–34; John E. Parson, *The First Winchester* (New York, 1955), 87–90; Kennett and Anderson, *The Gun in America*, 94–95.

97. Norton, *American Inventions and Improvements*, 71. Additional descriptions of the Peabody in battle may be found in the mammoth compendium of the Military-Historical Commission of the General Staff entitled *Sbornik materialov po russko-turetskoi voine 1877–1878 gg. na Balkanskom poluostrove*, 97 vols. (St. Petersburg, 1909–1911), 3: 272, 292–93, 325; 8: 4, 33; 33: 119; 53: 326, 328; 54: 63; 74: 27; 75: 180; 91: 102.

98. Argamakov, "Vospominaniia," 4: 77; 7: 149, 169, 166, 179.

99. *Ocherk preobrazovaniia*, 283.
100. Menning, *Bayonets Before Bullets*, 98. Verkhovskii, *Ocherk*, 224, 228; Strokov, *Istoriia voennogo iskusstva*, 573; *Small Arms of the World*, 68.
101. Argamakov, "Vospominaniia," 7: 160–61. Additional descriptions of the deployment of the Berdan and of its performance in battle may be found in *Sbornik materialov*, 1: 113–16; 11: 7–9; 31: 294, 353; 47: 251; 49: 299; 52: 159; 54: 134; 56: pt. 1, 41–42; 57: pt. 1, 3; 67: 358; 71: 10; 73: 192; 76: 14; 79: 84, 101; 90: 178; 91: 70, 155, 168, 175.
102. Wintringham, *The Story of Weapons*, 154.

Chapter 6. Labor, Management, and Technology Transfer

1. Quoted in Ashurkov, "K istorii," 58–59. See also V. N. Ashurkov, "Russkie oruzheinye zavody v 40-50kh gg. XIX v." in *Voprosy voennoi istorii Rossii* (Moscow, 1969), 207, 211–13.
2. Glebov, "Koe-chto," 175. It is perhaps not accidental that these words were penned in December 1861. Many foreigners, of course, shared this view. British artillery officers on an official mission in 1867 noted that the "soldiers" currently working at the armories were soon to be replaced by "civilian artisans, who presumably will work better" (N. O. S. Turner, F. G. E. Warren, and J. P. Nolan, *Tour of Artillery Officers in Russia* [London, 1867], 33).
3. "Doklad," quoted in Skalon, *Stoletie*, 1: 144.
4. Orfeev, "Istoriia Sestroretskogo oruzheinogo zavoda," 1–3; Aleksandrov, *Izhevskii zavod*, 95–105; *Istoriia Tul'skogo oruzheinogo zavoda*, 65–70.
5. Aleksandrov, *Izhevskii zavod*, 104–6. Although wages rose 67 percent after emanicpation, food prices rose 133 percent and the price of flour alone rose by 220 percent. The number of livestock fell and taxes increased.
6. Glebov, "Koe-chto" 190–91. The same rural-urban emigration patterns prevailed among peasants in the hinterland of Moscow. See my *Muzhik and Muscovite: Urbanization in Late-Imperial Russia* (Berkeley and Los Angeles, 1985), chapter 4.
7. V. N. Ashurkov, "Golos Tul'skikh oruzheinikov," *Katorga i ssylka 2*, 99 (1933): 141.
8. Ibid., 143.
9. P. A. Zaionchkovskii, *Voennye reformy*, 141. See also Glebov, "Koe-chto," 190–91.
10. Zaionchkovskii, *Voennye reformy*, 141; *Sestroretskii instrumental'nyi zavod*, 34.
11. *Istoriia Tul'skogo oruzheinogo zavoda*, 67–69; *Sestroretskii instrumental'nyi zavod*, 34–35; *Izhevskii zavod*, 95; Orfeev, "Istoriia Sestroretskogo oruzheinogo zavoda," 12, 21, 33.
12. Jacob W. Kipp, "The Russian Navy and the Problem of Technological Transfer: Technological Backwardness and Military-Industrial Development, 1853–1876," presented at the conference "The Great Reforms in Russian History," University of Pennyslvania, 25–28 May 1989.

13. Quoted in Zaionchkovskii, *Voennye reformy*, 142.
14. Miliutin, "Doklad," in Skalon, *Stoletie*, 1: 144; Maikov, "O proizvoditel'nykh silakh," 144; Graf, "Oruzheinye zavody," 388, 393; Chebyshev, "Po voprosu," 251–52.
15. Chebyshev, "Po voprosu," 251; M. Subbotkin, "Ob Izhevskom oruzheinom zavode," 177.
16. V. N. Zagoskin, "O tekhnicheskikh usloviakh deshevogo proizvodstva metallicheskikh patronnykh gil'z," *O.S.* 3 (1875): 29. Emphasis added.
17. Turner, *Tour*, 32.
18. S. Zybin, M. Nekliudov, M. Levitskii, *Oruzheinye zavody: Tul'skii, Izhevskii, Sestroretskii* (Kronstadt, 1898), 70.
19. Zagoskin, "O tekhnicheskikh usloviiakh," 30–31.
20. Chebyshev, "Po voprosu," 253.
21. Zagoskin, "O tekhnicheskikh usloviiakh," 32–33.
22. Quoted in Zaionchkovskii, *Voennye reformy*, 141.
23. Miliutin, "Doklad," in Skalon, *Stoletie*, 1: 145.
24. "Polozhenie o Tul'skom oruzheinom zavode," *O.S.* 3 (1870): 4–37; "Istoricheskii obzor Tul'skogo oruzheinogo zavoda," 6; Ashurkov, "K istorii promyshlennogo perevorota," 112–14; Ashurkov, *Kuznitsa oruzhiia*, 66–67.
25. Compare this with the inability of the Chinese hybrids—government supervised and privately managed—to perform the same function (Brown, "Transfer," 183).
26. Ashurkov, *Kuznitsa oruzhiia*, 67.
27. Chebyshev, "Po voprosu," 257; idem, "Ob ustroistve i naivygodneishem, s teoreticheskoi tochki zreniia, upotreblenii sharoshek," *O.S.* 1 (1875): 1.
28. "Perechen' zaniatii Oruzheinoi komissii," *A. Zh.* 9 (September 1861): 397–98; V. Bestuzhev-Riumin, "Ruchnoe ognestrel'noe oruzhie na parizhskoi mezhdunarodnoi vystavke," *O.S.* 4 (1867): 61; V. Chebyshev, "Peremena obraztsa dlia peredelki nashikh ruzhei," *O.S.* 2 (1869): 3.
29. Chebyshev, "Po voprosu," 250.
30. Chebyshev, "Peremena," 3; Chebyshev, "Po voprosu," 258–64; "Opyt teorii ruzheinogo dela," *O.S.* 1 (1861): 2. The impression that gunmakers worked better at home was shared by Subbotkin ("Ob Izhevskom oruzheinom zavode," 167) and Maikov ("O proizvoditel'nykh silakh," 60). Miasoedov, "Sravnitel'noe opisanie," 43, echoed Chebyshev's concerns about the great costs involved in systems changes.
31. Hounshell, *From the American System*, 48–49; Howard, "Interchangeable Parts," 642.
32. Eksten, *Opisanie*, 155.
33. Buniakovskii, "Neskol'ko slov," 3.
34. Ibid., 14; Buniakovskii, "Ustroistvo," 60. Colt's greatest suppliers of steel were Firth in Sheffield and Berger in Witten on the Ruhr in Westphalia. Russian steel was also judged to be of insufficient quality for high-powered rifle barrels. See A. N. Grodnitskii, "Fabrikatsiia bessemerovskoi stali dlia izgotovleniia voennogo oruzhiia," *Zapiski R. T. O.* 2 (1875): 25–26. The stan-

dard contemporary history of the Obukhov factory is V. Kolchak, *Istoriia Obukhovskogo staleliteinogo zavoda v sviazi s progressom artilleriiskoi tekhniki* (St. Petersburg, 1903). Unfortunately, Kolchak gives no attention to small arms production in this work.

35. Buniakovskii, "Neskol'ko slov," 14; Buniakovskii, "Ustroistvo," 60.
36. Buniakovskii, "Ustroistvo," 63, 64, 70.
37. A. Fon-der-Khoven, "Materialy dlia ocherka razvitiia ruzheinoi fabrikatsii v SShA," *O.S.* 4 (1877): 17–18, 22.
38. Buniakovskii, "Ustroistvo," 55–56, 58–59; P. Bil'derling, "Izgotovlenie stvolov na kol'tovskom oruzheinom zavode v Amerike," *O.S.* 4 (1870): 50; V. Eksten, *Opisanie*, 155. Bil'derling (p. 38, 45) also had praise for the vertical boring machines, no doubt the ones invented by Elisha Root, superintendent of the armory. On Root, see Joseph Wickham Roe, *English and American Tool-Builders* (New Haven, Conn., 1916), 169.
39. Quoted in Fedorov, *Vooruzhenie russkoi armii v XIX v.*, 245.
40. N. Litvinov, "Opisanie skorostrel'noi vintovki Berdana," *O.S.* 3 (1869): 109–10. Unpublished materials cited by Fedorov, (*Vooruzhenie russkoi armii v XIX v.*), provide additional support for this interpretation.
41. Buniakovskii, "Ustroistvo," 63; Miasoedov, "Sravnitel'noe opisanie," 74.
42. Berdan to Franklin, 28 September 1870, St. Petersburg, Box 4, Franklin Folder, CPFAM Records, RG 103, CSL.
43. Miasoedov, "Sravnitel'noe opisanie," 74.
44. *The Telegraph*, 14 February 1877, 3.
45. *Ocherk preobrazovaniia*, 309–10; "Perevooruzhenie nashei armii," *Russkii invalid* 260 (21 November 1871): 2–3.
46. "Bestuzhev-Riumin," *Voennaia entsiklopediia*, 4: 517; and "Notbek," *Voennaia entsiklopediia*, 17: 45.
47. Buniakovskii, "Ustroistvo," 57; *Ocherk preobrazovaniia*, 299–300.
48. Russian Government Contract, Order Books, Greenwood and Batley Records, Leeds District Archives, Leeds, England; Wellesley to Loftus, 3 May 1870 and 28 October 1873, St. Petersburg, Foreign Office Records, FO 802 and FO 519/277, PRO; Berdan to Franklin, 5 October and 6 November 1870, St. Petersburg, Box 4, Franklin Folder, CPFAM Records, RG 103, CSL; "Istoricheskii obzor Tul'skogo oruzheinogo zavoda," 11, 20–21; *Russkii invalid* 265 (26 November 1873): 2. Annual production was set at 75,000 rifles ("Perevooruzhenie Evropy," *O.S.* 3 [1872] 41–45).
49. Thomas Greenwood to James Henry Burton, 19 March and 25 April 1871, Leeds, James Henry Burton Papers, vol. 2, Yale University Library. Several of the new machines were of American design, including seven American-pattern rifling machines, fifteen Blanchard lathes, and two American-pattern universal milling machines. (Russian Government Contract, Order Books, Greenwood and Batley Records, Leeds District Archive.)
50. A. Druzhinin and N. P. Mironov, "Mashinnoe izgotovlenie malokalibernykh vintovok v Tul'skom oruzheinom zavode," *O.S.* 4 (1875): 1–13; Miasoedov, "Sravnitel'noe opisanie," 76; Ashurkov, *Kuznitsa oruzhiia*, 68; *Moskovskaia promyshlennaia vystavka s voennoi tochki zreniia* (St. Pe-

tersburg, 1872), 93; *Otchet po venskoi vsemirnoi vystavke,* 235–36. Little information on Burton in Russia exists. His papers, at Yale University Library, are reticent. A paper on Burton by Edward Ezell prepared for the Eleutherian Mills-Hagley Foundation, and cited in the introduction to the English translation of Gamel', was unavailable to me.

51. "Istoricheskii obzor Tul'skogo oruzheinogo zavoda," 6–8; Ashurkov, *Kuznitsa oruzhiia,* 68–70.

52. Ibid; "Notbek," *Voennaia entsiklopediia,* 17: 45.

53. S. Zybin, et al., *Oruzheinye zavody,* 28–29; "Pravitel'stvennye rasporiazheniia v prikazakh po artillerii, no. 40," *O.S.* 3 (1872): 19–25; I. A. Krylov, *Opisanie Imperatorskogo Tul'skogo oruzheinogo zavoda* (Tula, 1903), 8; V. N. Ashurkov, *Tul'skii muzei oruzhiia* (Tula, 1972): V. N. Ashurkov, *Russkie oruzheinye zavody vo II polovine XIX v.,* Candidate's dissertation (Moscow, 1962), 22; G. G. Zashchuk, *Malokalibernaia skorostrel'naia vintovka Berdana N. 2* (St. Petersburg, 1872), 4.

54. Conversation with Runo Kurko, a Finnish historian of small arms, in Helsinki, June 1979. See also "Perechen' zaniatii oruzheinogo otdela artilleriiskogo komiteta," *O.S.* 2 (1886): 56; Ashurkov, *Tul'skii muzei oruzhiia,* 38; Ashurkov, "Oruzheinye zavody," 23; Markevich *Ruchnoe ognestrel'noe oruzhie,* 397.

55. *Moskovskaia promyshlennaia vystavka,* 93; Ashurkov, *Tul'skii muzei oruzhiia,* 38–40. Likewise a model workshop was set up to fill special orders, especially of revolvers. See Iu. V. Shokarev, "Khudozhestvennoe oruzhie Tul'skogo oruzheinogo zavoda," *Soobshcheniia Ermitazha* 41 (1976): 305–10.

56. Wellesley to Loftus, no. 66, 28 October 1873, St. Petersburg, Foreign Office Records, FO 519/277, PRO.

57. *Moskovskaia promyshlennaia vystavka,* 38–40. See also *Russkii invalid* 266 (1873): 4.

58. *The Industries of Russia,* 97. Repeatedly in Russian descriptions of Tula and Izhevsk firearms one notices the pride of unexpected accomplishment, particularly if such firearms were on display at international expostions and noticed by foreigners. Bestuzhev-Riumin's reaction was typical when he observed the display of Tula, Izhevsk, and Sestroretsk firearms at the Paris Exposition of 1867: "We Russians are so used to our backward factories making inferior products and to ourselves bowing before the work of foreign craftsmen, that many will be surprised by the impression that these models made on foreigners and on the judges. . . . The work of our factories surprised everybody: nothing like it had been seen before at the government armories of foreign countries" ("Ruchnoe oruzhie," 48).

59. Ashurkov, *Kuznitsa oruzhiia,* 69–70. This estimation must be tempered not only by the fact that Ashurkov was writing in 1947, a time when Russian feats were exaggerated, but also by the fact that this claim was made for Tula in the 1820s, as discussed in Chapter 4.

60. Prerevolutionary histories of the two armories may be found in Subbotkin, "Ob Izevskom oruzheinom zavode"; Maikov, "O proizvoditel'nykh silakh"; A. Solov'ev, "Materialy dlia istorii Izhevskogo zavoda," *O.S.* 4 installments (1902–1904); Orfeev, "Istoriia Sestroretskogo zavoda"; Graf, "Oruzheinye zavody"; and Zybin, *Oruzheinye zavody*.

61. "Istoriia Sestroretskogo zavoda," 18, 21; "Perechen' zaniatii Komiteta ob uluchshenii shtutserov i ruzhei za ianvar' i fevral'," *A. Zh.* 1 (1857): 67–68; Evgenii Gutor, *Amerikanskie mashiny dlia vydelki ruzheinykh lozh* (St. Petersburg, 1868). O. F. Lilienfel'd became superintendent at Sestroretsk in 1861. He had begun his technical training at Izhevsk, where he studied barrel making, and "brilliantly continued it in America." See O. F. Lilienfel'd, "Mashinnaia i ruchnaia zavarka stvolov," *A. Zh.* 12 (1861): 991–1022.

62. Orfeev, "Istoriia Sestroretskogo zavoda," 8; *Kratkii obzor preobrazovanii po artillerii* (St. Petersburg, 1863), 52–53; Gutor, *Amerikanskie mashiny*, 2, 34; *Sestroretskii instrumental'nyi zavod*, 31, 35. The 34 machines Colt sold on 1 October 1856 were installed at Sestroretsk. The machines would have been in operation as early as 1858 had they not been for a large caliber rifle. Adjusting them to the Russian muzzleloader, recently adotped, took an additional two years.

63. Turner, *Tour*, 64. The Ames Company also sold machinery to the British government when the Royal Enfield Armory was equipped with American machinery after the Crimean War (see Rosenberg, *The American System*, Introduction, passim). Some American lock-making machinery was also introduced, resulting in partial mechanization of this part. These machines included a screw press and a vertical milling machine. See "Opisanie razrabotok, upotrebliaiushchikhsia na Sestroretskom oruzheinom zavode dlia izgotovleniia zamochnykh chastei nashei 6-lineinoi vintovki," *O.S.* 1 (1869): 23.

64. Gutor, *Amerikanskie mashiny*, vi, 35. Judges at the Paris Exposition in 1867 had high expectations for the Sestroretsk entries because they had heard that "American stockmaking machines had been introduced at Sestroretsk and such were few in Europe" (Bestuzhev-Riumin, "Ruchnoe oruzhie," 51).

65. Zybin, et al., *Oruzheinye zavody*, 52–53.

66. V. Bestuzhev-Riumin, "Neskol'ko slov o vvedenii u nas litoi stali dlia ruzheinykh stvolov," *O.S.* 1 (1863): 141–46; V. I. Sirota, "Perevooruzhenie russkoi armii vo II poloviny XIX v.," Candidate's dissertation (Leningrad, 1950), 15.

67. Aleksandrov, *Izhevskii zavod*, 57–61. According to the jubilee factory history, 65 percent of iron barrels made from 1842 to 1855 were defective Solov'ev, *V pamiat'*, 28.

68. Ashurkov, "Oruzheinye zavody," 22.

69. Aleksandrov, *Izhevskii zavod*, 108; Zaionchkovskii, *Voennye reformy*, 143; *Izhevsk, 1760–1985: Dokumenty i materialy* (Izhevsk, 1984), 41. Steel barrels were made on a milling machine that, according to Edwin A. Batti-

son, a specialist in the history of machines, bore a strong resemblance to the Howe and Springfield milling machines. However, with the evidence available, it is impossible to determine whether the milling machine was a foreign copy or a Russian invention. See Battison's Introduction to the Smithsonian Institution's translation of Zagorskii, *Ocherki*, 11.

70. Aleksandrov, *Izhevskii zavod*, 31, 34.

71. Ibid., 32, 36–37, 58–59; Lilienfel'd, "Zavarka stvolov," 995; "Izhevskii oruzheinyi zavod," *Russkii invalid* 263 (1873): 3.

72. P. Kharinskii, "Liteinyi gil'zovyi otdel patronnogo zavoda," *O.S.*1 (1883): 3–4; A. Pelekin, "Fabrikatsiia tsel'no-tianutykh metallicheskikh patronov sistemy Berdana," *O.S.* 1 (1871): 33; "Opisanie ustroistva i chertezh metallicheskikh gil'z sistemy Berdana k patronam dlia shestilineinykh skorostrel'nykh udarnykh vintovok Krynka i kratkoe opisanie ustanovivsheisia u nas fabrikatsii ikh," *O.S.* 2 (1870): 1; "Deiatel'nost' Russkogo Tekhnicheskogo obshchestva v artilleriiskom otnoshenii," *A. Zh.* 4 (1873): 482–95.

73. According to Miliutin, cited in Askew, "Russian Military Stregth," 193, when the Russians began to manufacture metallic cartridges in 1868, several Americans were employed, and one shop was even put under the direction of a person named A. Sexton. Later, when Russian artillery officers were put in charge and Sexton became "bogged down by red tape," daily production dropped from 20,000 to 5,000. I have found no confirmation of this in any other source.

74. Norton, *American Inventions and Improvements*, 309–10; Kharinskii, "Liteinyi otdel," 5; Gnatkovskii, *Istoriia*, 109. Norton claimed that the Coe Brass Company sold cartridges "in enormous quantities for use, on both sides, in the Russo-Turkish War" (p. 310). American-made powder also turned out to be of higher quality than Russian-made. See A. Slukhinskii, "O novom ruzheinom porokhe dlia malokalibernoi vintovki," *O.S.* 1 (1878): 32.

75. A. A. Korolev, "Iz istorii russkoi voennoi promyshlennosti: Vozniknovenie Tul'skogo chastnogo patronnogo zavoda," *Istoricheskie nauki: Uchenye zapiski kafedry istorii* 2 (Tula, 1969): 88–98.

76. P. Kharinskii, "Opyty nad sistemoi malokalibernykh gil'z," *O.S.* 3 (1880): 1.

77. Kharinskii, "Liteinyi otdel," 3–5; Mavrodin and Mavrodin, *Iz istorii*, 58. Chebyshev also argued that production had to be set up gradually as workers were trained to use one series of machines at a time ("Opisanie ustroistva metallicheskikh patronov, priniatykh dlia voennogo oruzhiia i posledovatel'nogo ikh usovershenstvovaniia," *O.S.* 2 [1871]: 18).

78. Zagoskin, "O vvedenii," 157. See also "Deiatel'nost' Russkogo Tekhnicheskogo obshchestva," 485.

79. *The Industries of Russia*, 136.

80. "Tul'skii oruzheinyi zavod," 111–12. Small wonder that eight years later, Iosif Gamel' attributed publication difficulties encountered in his book on Tula to "the lack of draftsmen and engravers in this country" (*Opisanie*, xiii).

81. *Sestroretskii instrumental'nyi zavod*, 27–29, 35. The first proposal for a tool shop in the small arms industry actually came at the opening of the Izhevsk armory in 1808. However, it was 90 years before a tool shop opened at Izhevsk (Solov'ev, *V pamiat'*, 45).

82. "Istoricheskii obzor," 11, 30; Goldstein, "The Military Aspects of Russian Industrialization," 90.

83. Hogan, "Labor and Management in Conflict," 14, 42–43.

84. The case of Russia's production of the Maxim machine gun comes from E. V. Myshkovskii, "Russkoe avtomaticheskoe oruzhie: Rabota russkikh oruzheinikov, 1887–1917 gg.," *Sbornik issledovanii i materialov Artilleriiskogo istoricheskogo muzeia*, 3 (Leningrad, 1958): 194–203.

85. Godfrey L. Carden, *The Machine Tool Trade in Austria-Hungary, Denmark, Russia and the Netherlands*, Department of Commerce and Labor, Special Agent Series, no. 34 (Washington, D.C., 1910), 119–20, 130–31. The successful example Carden had in mind was the Pneumatic Tool Company of St. Petersburg, whose head was John K. Lemke, a native American.

86. Ashurkov, "Russkie oruzheinye zavody," 118–20, 128; Beskrovnyi, *Russkaia armiia i flot*, 315. The standard Soviet history of Leningrad reaches the same conclusion regarding the machine-building industry, which imported many machines. See Akademiia nauk SSSR, *Ocherki istorii Leningrada*, vol. 2 (Moscow, 1957), 107.

87. Ashurkov, *Kuznitsa oruzhiia*, 73; *Istoriia Sestroretskogo instrumental'nogo zavoda*, 35; *Istoriia Tul'skogo oruzheinogo zavoda*, 84–85; "Mosin," *Voennaia entsiklopediia*, 16: 444.

88. Ashurkov, "Russkie oruzheinye zavody," 114, 119, 129; Gorbov, *Izhevskie oruzheiniki*, 17.

89. V. A. Tsybul'skii, "Sestroretskii oruzheinyi zavod i perevooruzhenie russkoi armii v kontse XIX v.," in V. V. Mavrodin, ed., *Rabochie oruzheinoi promyshlennosti v Rossii i russkie oruzheiniki v XIX-nachale XX v.* (Leningrad, 1976), 62–64; Ashurkov, "Russkie oruzheinye zavody," 118–19, 122; Beskrovnyi, *Russkaia armiia i flot*, 315.

90. Beskrovnyi, *Russkaia armiia i flot*, 319; Gorbov, *Izhevskie oruzheiniki*, 17. Izhevsk had the greatest problem in retooling to make the Mosins and as late as 1895 was still not completely mechanized.

91. Beskrovnyi, *Russkaia armiia i flot*, 319–21, 323.

92. Ashurkov, "Russkie oruzheinye zavody," 123.

93. Tsybul'skii, "Sestroretskii oruzheinyi zavod," 64. Emphasis added. Not only was the pattern of foreign orders repeated in the arms industry, it was also repeated in other branches of industry. For the case of the chemical industry, see Kirchner, "Russian Entrepreneurship," 82.

94. Ashurkov, "Russkie oruzheinye zavody," 129.

95. Miliutin, "Doklad," in Skalon, *Stoletie*, 1: 145.

96. In addition to Miliutin, see, for example, Turner, *Tour*, 33; and Barry, *Russian Metallurgical Works*, 37.

97. Chebyshev, "Po voprosu," 250.

98. "Istoricheskii obzor Tul'skogo oruzheinogo zavoda," *O.S.* 4 (1873): 4–5.

99. The phrase is Berdan's in a letter to William Franklin describing a conversation with General Notbek, the commander of the Tula armory, in which Berdan tried to convince the Russians to order machines from Colt (Berdan to Franklin, 28 September 1870, St. Petersburg, Box 4, Franklin Folder, CPFAM Records, RG 103, CSL).

Chapter 7. Conclusion

1. Fedorov, *Evoliutsiia*, 6.
2. Lewis Mumford, *Technics and Civilization*, 2d ed. (New York, 1963), 95; Showalter, *Railroads and Rifles*, 12–13, 163.
3. Fedorov, *Vooruzhenie russkoi armii v XIX v.*, 238.
4. Miliutin, "Doklad," in Skalon, *Stoletie*, 1: 135–36; Fedorov, *Vooruzhenie russkoi armii v XIX v.*, 120.
5. Chebyshev, "Po voprosu," 250.
6. Turner, *Tour*, 32. See also "Istoricheskii obzor," 5–6.
7. Subbotkin, "Ob Izhevskom oruzheinom zavode," 159; Glebov, "Koe-chto o tul'skikh oruzheinikakh," 173–74.
8. For example, see Bonnell, *The Russian Worker* and *Roots of Rebellion*, 62–67; Zelnik, *Radical Worker*; Hogan, "Labor and Management," 164–65; Hogan, "Industrial Rationalization," 165–66.
9. Departament manufaktur i vnutrennei torgovli, *Obzor razlichnykh otraslei manufakturnoi promyshlennosti Rossii*, 3 vols. (St. Petersburg, 1862), 2: 81; Orfeev, "Istoriia Sestroretskogo oruzheinogo zavoda," 30–31.
10. *Obzor razlichnykh otraslei*, 2: 81. See also "Istoricheskii obzor," 30. For similar comments by representatives of International Harvester in Russia, see Carstensen, *American Enterprise*, 51, 71.
11. Kaser, "Russian Entrepreneurship," 424–30. On the "culture of productivity," see Ashurkov, *Kuznitsa*, 70.
12. Chebyshev, "Po voprosu," 253.
13. Ashurkov, *Kuznitsa*, 67.
14. Granick, *Soviet Metal-fabricating*, 24.
15. Ibid., 25, 112. See also Kendall E. Bailes, *Technology and Society under Lenin and Stalin: Origins of the Soviet Technical Intelligentsia, 1917–1941* (Princeton, 1978).
16. Philip Hanson, *Trade and Technology in Soviet-Western Relations* (New York, 1981), 186; Eugene Zaleski and Helgand Wienert, *Technology Transfer between East and West* (Paris, 1980), 214.
17. Brada, "Soviet-Western Trade," 23; Kaser, "Russian Entrepreneurship," 422.
18. Hanson, *Trade and Technology*, 63, 195; Zaleski and Wienert, *Technology Transfer*, 195, 198–99.
19. Hanson, *Trade and Technology*, 43. See also Zaleski and Wienert, *Technology Transfer*, 198–99; R. Amann, J. M. Cooper, R. W. Davies, eds., *The Technological Level of Soviet Industry* (New Haven, 1977), 58. The best overall study of innovation in the Soviet economy is Joseph Berliner, *The Innovation Decision in Soviet Industry* (Cambridge, Mass., 1978).

20. Hanson, *Trade and Technology*, 55, 63; Zaleski and Wienert, *Technology Transfer*, 198–99.
21. Amann, et al., *Technological Level*, 58, 160–64, 194; Hanson, *Trade and Technology*, 199.
22. Amann, et al, *Technological Level*, 414–15, 438; Ronald Amann, "Technological Progress and Soviet Economic Development: Setting the Scene," in Ronald Amann and Julian Cooper, eds., *Technical Progress and Soviet Economic Development* (New York, 1986), 15; Dale R. Herspring, "Technology and the Soviet System," *Problems of Communism* (January–February 1985), 73–76.
23. Gary K. Bertsch, "Technology Transfers and Technology Controls: A Synthesis of the Western-Soviet Relationship," in Amann and Cooper, eds., *Technological Progress*, 124–26; Brada, "Soviet-Western Trade," 30. This is also the conclusion of A. C. Sutton, *Western Technology and Soviet Economic Development*, 3 vols. (Stanford, 1968, 1971, 1973).

Bibliography

Archival Sources

Burton, James H. Papers. Yale University Library. New Haven, Conn.

Colt Patent Fire Arms Manufacturing Company Records, 1835–1968. Record Group 103. Connecticut State Library. Hartford, Conn.

Colt, Samuel. Papers. Connecticut Historical Society. Hartford, Conn.

Colt, Samuel. Papers. Wadsworth Atheneum. Hartford, Conn.

Department of State Foreign Service Posts. Record Group 84. National Archives. Washington, D.C.

Department of State Papers. Record Group 59. National Archives. Washington, D.C.

Foreign Office Records. FO 65. Public Records Office. London.

Greenwood and Batley Records. Leeds District Archives. Leeds, England.

Smith and Wesson Records. Private Collection of Roy Jinks. Springfield, Mass.

Primary Sources

Aleksandrov, M. "Igol'chataia skorostrel'naia vintovka Karle." (The Carl needle gun). *O.S.* 1, 2 (1876): 1–27.

——. *Pekhotnaia malokalibernaia vintovka so skol'ziashchim zatvorom* (The sliding bolt action small-caliber infantry rifle). St. Petersburg, 1876.

——. *Revol'ver Smita i Vesona 2-go obraztsa dlia kavalerii* (The model 2 Smith and Wesson cavalry revolver). St. Petersburg, 1876.

Alekseev, N. "Metallicheskii malokalibernyi patron" (The small-caliber metallic cartridge). *O.S.* 1 (1877): 1–36.

Anquetil, Thomas. *Notice sur les pistolets tournants et roulents, dits revolvers, ou leur passé, leur présent, leur avenir, suivié des principes generaux sur le tir de ces armes* (A description of the turning and revolving pistols, called revolvers, their past, present, and future, with a survey of the general principles for firing these weapons). Paris, 1854.

Argamakov, V. F. "Vospominaniia o voine 1877–1878 gg." (Reminiscences of the War of 1877–1878) *Zhurnal Imperatorskogo Russkogo Voenno-istoricheskogo obshchestva* 2–7 (1911): 1–184.

Astaf'ev, A. I. *O sovremennom voennom iskusstve* (The art of war today). St. Petersburg, 1856.

Barnard, H. *Armsmear: The Armory of Samuel Colt.* Boston, 1866.

Barry, Herbert. *Russian Metallurgical Works, Iron, Copper and Gold, Concisely Described.* London, 1870.

Bestuzhev, I. V., and D. A. Miliutin. "Ob opasnosti prodolzheniia v 1856 g. voennykh deistvii" (The danger of continuing the war in 1856). *Istoricheskii Arkhiv* 1 (1959): 204–8.

Bestuzhev-Riumin, V. "Neskol'ko slov o vvedenii u nas litoi stali dlia ruzheinykh stvolov" (A comment on the introduction of steel for our gun barrels). *O.S.* 1 (1863): 141–46.

———. "Razbor i opisanie izgotovleniia shestilineinoi vintovki na Sestroretskom oruzheinom zavode" (Description and analysis of the manufacture of the .60-caliber rifle at the Sestroretsk armory) *A. Zh.* 7 (1857): 17–29.

Bil'derling, P. "Izgotovlenie stvolov na kol'tovskom oruzheinom zavode v Amerike" (Barrel manufacture at the Colt armory in America). *O.S.* 4 (1870): 32–57.

Bil'derling, P., and V. Buniakovskii, "Russkaia igol'chataia vintovka" (The Russian needle gun). *O.S.* 1 (1868): 1–70; 2 (1868): 23–48; 4 (1868): 15–37; 1 (1869): 1–65; 2 (1869): 1–9.

Bogdanovich, K. I., comp. *Istoricheskii ocherk deiatel'nosti Voennogo upravleniia v Rossii v pervoe 25-letie blagopoluchnogo tsarstvovaniia Gosudaria Imp. Aleksandra Nikolaevicha, 1855–1880* (The history of the Russian war office during the first 25 years of the glorious reign of Aleksandr II, 1855–1880). 6 vols. St. Petersburg, 1879–1881.

Bogdanovich, M. I. "O noveishikh usovershenstvovaniiakh ruchnogo ognestrel'nogo oruzhiia" (The latest improvements in firearms). *Voennyi zhurnal* 3 (1854).

Buniakovskii, V. "Neskol'ko slov o svoistvakh russkoi .42-lineinoi vintovki, ob ispytanii onoi v Amerike, ob uluchsheniiakh proizvedennykh v nei i o preimushchestvakh onoi pered drugimi obraztsami oruzhiia, zariazhaiushchegosia s kazni" (A comment on the features of the .42-caliber Russian rifle, on its tests in America, on the improvements made, and on its advantages over other breechloading rifles). *O.S.* 4 (1869): 1–25.

———. "O metallicheskikh patronakh k russkoi malokalibernym vintovkam" (Metallic cartridges for the Russian small-caliber rifles). *O.S.* 2 (1870): 53–70.

———. "Ustroistvo Kol'tovskogo oruzheinogo zavoda" (The organization of the Colt armory). *O.S.* 4 (1869): 53–79.

Chebyshev, V. "O revol'verakh i okhotnich'ikh ruzh'iakh dlia ofitserov" (Revolvers and hunting arms for officers). *O.S.* 3 (1872): 29–33.

———. "O sovremennom polozhenii patronnogo voprosa" (The current state of the cartridge problem). *Zapiski R. T. O.* 1 (1881): 32–42.

———. "O sredstvakh k umen'sheniiu poter' ot ognia pri atake pekhoty, s tochki zreniia svoistv ruzheinogo ognia" (Ways to decrease losses from infantry attack fire, considering the capabilities of small arms fire). *V.S.* 10 (1878): 209–40.

———. "Opisanie ustroistva metallicheskikh patronakh, priniatykh dlia voennogo oruzhiia i posledovatel'nogo ikh usovershenstvovaniia" (A description of the metallic cartridges adopted for military arms and of their subsequent improvements). *O.S.* 1 and 2 (1871): 1–14 and 1–69; 1 and 2 (1872): 29–69 and 77–87.

———. "Peremena obraztsa dlia peredelki nashikh ruzhei" (The change in the conversion model for our small arms). *O.S.* 2 (1869).

———. "Po voprosu o novom administrativnom i tekhnicheskom ustroistve Tul'skogo oruzheinogo zavoda" (The new administrative and technical organization of the Tula armory). *V.S.* 12 (1869): 249–74.

"Colt and His Revolvers." *Colburn's United Service Magazine* 1 (1854): 118–21.

Colt, Samuel. "On the application of machinery to the manufacture of rotating chambered-breech firearms and the peculiarities of those arms." *Minutes of the Proceedings of the Institute of Civil Engineers* 11 (1851–1852): 30–68.

Colt's Patent Fire Arms Manufacturing Company: A Century of Achievement, 1836–1936. Hartford, 1937.

D. [Dragomirov, M.] "O veroiatnykh peremenakh v taktike vsledstvie rasprostraneniia dal'no- i skorstrel'nogo oruzhiia" (Probable changes in tactics resulting from the spread of long-range and rapid-fire small arms). *V.S.* 11 (1867).

Departament manufaktur i vnutrennei torgovli (Department of manufacturing and internal commerce). *Obzor razlichnykh otraslei manufakturnoi promyshlennosti Rossii* (A survey of various branches of manufacturing in Russia). 3 vols. St. Petersburg, 1862.

DeSharer, G. *Kratkie zametki o skorostrel'nom ruchnom oruzhii* (Notes on firearms). St. Petersburg, 1870.

Dragomirov, M.I. "O sredstvakh, sposobstvuiushchikh razvitiiu takticheskikh poznanii v voiskakh" (The methods that develop an understanding of tactics in soldiers). *O.S.* 1, 4 (1862).

———. *Sbornik original'nykh i peredovykh statei* (A collection of original and translated articles). 2 vols. St. Petersburg, 1881.

———. *Uchebnik taktiki* (Tactics). St. Petersburg, 1881.

———. "Vintovki, skorostrel'naia i obyknovennaia, s takticheskoi tochki zreniia" (Rapid-fire and conventional rifles from the tactical point of view). *A. Zh.* 3 (1869): 463–96.

———. "Vliianie rasprostraneniia nareznogo oruzhiia na vospitanii i taktiku voisk" (The effect of the spread of rifles on military training and tactics). *O.S.* 1 (1861): 48–95.

Druzhinin, A., and N. P. Mironov. "Mashinnoe izgotovlenie malokaibernykh vintovok v Tul'skom oruzheinom zavode" (Machine manufacture of small-caliber rifles at the Tula armory). *O.S.*, 12 installments (1875–1881).

Egershtrom, N. "Svedeniia, otnosiashchiesia do vvedeniia v russkoi armii ruchnogo oruzhiia umen'shennogo kalibra" (Information regarding the introduction of small-caliber firearms in the Russian army). *O.S.* 1 (1861): 21–49.

Eksten, V. *Opisanie sistem skorostrel'nogo oruzhiia* (A description of firearms' systems). Moscow, 1870.

Engel'gardt, A. P., and Litvinov, N. "Nastavlenie dlia peresnariazheniia metallicheskikh patronov k malokalibernomu oruzhiiu i skorostrel'nym pushkam" (A manual for refiring metallic cartridges for small-caliber firearms and rapid-fire cannons). *O.S.* 3 and 4 (1871).

Epikhin, M. *Sbornik postanovlenii o boevykh pripasakh k ognestrel'nomu ruchnomu oruzhiiu* (A collection of the regulations pertaining to firearm ammunition). St. Petersburg, 1863.

Fitch, Charles H. U.S. Census Office. *Report on the Manufacture of Fire-Arms and Ammunition.* Washington, D.C., 1882.

Fon-der-Khoven, A. "Kratkie svedeniia o proizvodstve laboratornykh rabot, fabrikatsii metallicheskikh patronov i merakh, prinimaemykh v pravitel'stvennykh zavodakh SShA, dlia umen'sheniia poter' v liudiakh, pri vzryvakh i pozharakh" (A note on laboratory experiments, on the fabrication of metallic cartridges, and on measures taken at government armories in the USA to reduce casualties caused by explosions and fires). *O.S.* 3 (1881): 1–39.

———. "Materialy dlia ocherka razvitiia ruzheinoi fabrikatsii v SShA" (A study of the development of arms fabrication in the USA). *O.S.* 4 (1877): 13–39.

———. "Zametki o venskoi vsemirnoi vystavke 1873 g." (Notes on the Vienna World Exposition of 1873). *O.S.*, 5 installments (1873–1874).

G., I. "Samuel' Kol't" (Samuel Colt). *A. Zh.* 1 (1868): 128–31.

Gamel', Iosif. *Opisanie Tul'skogo oruzheinogo zavoda v istoricheskom i tekhnicheskom otnoshenii* (A historical and technical description of the Tula small arms factory). Moscow, 1826.

Glebov, P. "Koe-chto o tul'skikh oruzheinikakh" (Regarding the Tula armorers). *A. Zh.* 2 (1862): 161–92.

Gol'mdorf, M., ed. *Ukazatel'-sbornik rasporiazhenii po artillerii za 20 let* (An indexed collection of artillery orders for the past 20 years). St. Petersburg, 1875.

Gorlov, Aleksandr. "O dvizhenii snariada v kanale nareznogo oruzhiia" (The movement of a projectile in a rifle barrel). *A. Zh.* 5 and 11 (1862): 453–70, 939–55.

———. "O povtoritel'nom ognestrel'nom oruzhii ili revol'verakh" (Rotating firearms or revolvers). *M.S.* 2 (1856): 426–59.

———. "Ob upotrebliaemykh v armii Soedinennykh Amerikanskikh Statov ruzh'iakh, zariazhaiushchikhsia s kazennoi chasti, i k nim metallicheskikh patronov" (The use of breechloading firearms with metallic cartridges in the United States Army). *O.S.* 3 (1866): 1–48; reprinted in *A. Zh.* 9 (1866).

Graf, F. "Oruzheinye zavody v Rossii" (Small arms factories in Russia). *V.S.* 9 and 10 (1861): 113–36, 365–94.

Great Exhibition of the Works of the Industry of All Nations. Official Description and Illustrated Catalogue. 3 vols. London, 1851.

Gutor, Evgenii. *Amerikanskie mashiny dlia vydelki ruzheinykh lozh:*

Opisanie mashinnogo izgotovleniia lozh na sestroretskom oruzheinom zavode (American stockmaking machines: a description of making stocks by machine at the Sestroretsk armory). St. Petersburg, 1868. Reprinted from *O.S.*, 6 installments (1865–1868).

"Instruktsiia, vremennaia dlia priema shestilineinykh kapsiul'nyk skorostrel'nykh vintovok na zavodakh" (Temporary instructions for the inspection of .60-caliber percussion rifles at factories). *O.S.* 3 (1867): 1–9.

International Exhibition of 1862. *Catalogue of the Russian Section*. London, 1862.

"Istoricheskii obzor Tul'skogo oruzheinogo zavoda i nastoiashchee ego polozhenie" (The history and current state of the Tula armory). *O.S.* 4 (1873): 1–31.

"Izmeneniia v shestilineinoi vintovke (Sestroretskim zavodom)" (Changes in the .60-caliber rifle at the Sestroretsk armory). *A. Zh.* 7 (1861): 316–22.

Kazantsev, M. "Pochinki malokalibernykh skorostrel'nykh strelkovykh vintovok obraztsa 1868 g." (Repairing the 1868 model small-caliber rifle). *O.S.* 1–3 (1872): 1–27, 1–39, 1–47.

Kharinskii, P. "Liteinyi gil'zovyi otdel patronnogo zavoda" (The casing shop of the cartridge factory). *O.S.* 1 (1883): 1–54.

———. "Opyty nad sistemoi malokalibernykh gil'z so sploshnoi golovkoi i fabrikatsiia gil'z amerikanskoi sistemy so vnutrennei chashkoi" (Tests of the small-caliber capped cartridge cases and the fabrication of inside cupped cartridge cases by the American system). *O.S.*, 13 installments (1874–1882).

Konstantinov, K. "Posledovatel'nye usovershenstvovaniia ruchnogo ognestrel'nogo oruzhiia" (Sequential improvements in firearms). *M.S.* 5 (1855): 1–53.

Korostovtsev, Col. A. "Obzor issledovaniia proizvedennykh u nas nad noveishimi sistemami ruchnogo ognestrel'nogo oruzhiia" (A survey of our research on the newest firearms' systems). *A. Zh.* 1 (1854): 1–64.

Kostenkov, K. *Opisanie revol'verov i pravila obrashchat'sia s nimi* (A description of revolvers and the rules for handling them). St. Petersburg, 1855.

Leer, A. "Takticheskie voprosy: Sredstva k umen'sheniiu poter' ot ogni pri atake pekhoty" (Tactical problems: ways to reduce losses from infantry attack fire). *V.S.* 2 and 4 (1878): 253–73; 228–52.

Lilienfel'd, Otto. "Mashinnaia i ruchnaia zavarka stvolov" (Machine- and hand-forged barrels). *A. Zh.* 12 (1861): 991–1022.

Litvinov, N. "Nasha peredelochnaia shestilineinaia vintovka po sisteme Krnka i patron Berdana" (Our Krnka .60-caliber conversion rifle with the Berdan cartridge). *R.I.* 132 (6/18 November 1869): 3.

———. "O poslednikh obraztsakh ruzhei i o metallicheskikh patronakh, sushchestviushchikh v nashei armii" (The latest model small arms and metallic cartridges in our army). *R.I.* 75, 79, 81 (1875).

———. "Opisanie skorostrel'noi vintovki Berdana" (A description of the Berdan rifle). *O.S.* 3 (1869): 14–31.

Lukin, P. "Novye obraztsy oruzhiia, zariazhaiushchegosia s kazni" (The new model breechloading small arms). *O.S.* 1 (1865): 1–17.

————. "Pochinka ruchnogo ognestrel'nogo oruzhiia v polkakh i masterskikh artilleriiskogo vedomstva" (Repairing firearms in the regiments and in shops of the artillery office). *O.S.*, 6 installments (1861–1868).

Maikov, P. M. "O proizvoditel'nykh silakh oruzheinykh zavodov Izhevskogo, Tul'skogo i Sestroretskogo" (The productive capacity of the Izhevsk, Tula, and Sestroretsk armories). *A. Zh.* 8 and 9 (1861): 583–613, 615–42; 1 and 2 (1862): 38–77, 122–54.

Majendie, V. D. "Military Breech-loading Small Arms." *Journal of the Royal United Services Institution* 11, 44 (1867): 190.

"Mashinnoe izgotovlenie malokalibernnoi vintovki v Tul'skom oruzheinom zavode" (Machine manufacture of the small-caliber rifle at the Tula armory). *O.S.* 4 (1875): 1–13.

"Memoriia artilleriiskogo upravleniia SShA ob ispytanii skorostrel'nykh ruzhei" (A notice of the U.S. Artillery Office on the tests of firearms). *O.S.* 3, 4 (1871): 1–22, 1–19.

Miasoedov, I. "Sravnitel'noe opisanie ruzhei Berdana 2, Mauzera, Gra i Gochkisa" (A comparison of the Berdan 2, Mauser, Ger, and Hotchkiss Rifles). *O.S.* 3 (1875): 27–82.

Miliutin, D. A. *Dnevnik* (Diary). 4 vols. Moscow, 1947–1950.

Moskovskaia politekhnicheskaia vystavka (The Moscow Exposition of Science and Industry). *Moskovskaia politekhnicheskaia vystavka s voennoi tochki zreniia* (The military aspects of the Moscow Exposition of Science and Industry). St. Petersburg, 1872.

————. *Ukazatel' kollektsii artilleriiskogo otdela Moskovskoi politekhnicheskoi vystavki 1872 g.* (Guide to the collection of the artillery section of the Moscow Exposition of Science and Industry). Moscow, 1872.

Naperstochnyi patron dlia ruzhei sistemy Berdana (The ring cartridge for the Berdan firearms). St. Petersburg, 1877.

"Nekotorye dannye dlia sravnitel'noi otsenki mashinnoi i ruchnoi vydelki lozh" (Some information for a comparison of machine- and hand-made gun stocks). *O.S.* 1 (1867): 49–76.

"Neskol'ko slov o znachenii ognestrel'nogo oruzhiia v kavalerii" (A note on the significance of firearms in the cavalry). *V.S.* 10 (1862): 375–84.

Norton, Charles. *American Inventions and Improvements in Breech-loading Small Arms, Heavy Ordnance, Machine Guns, Magazine Arms, Fixed Ammunition, Pistols, Projectiles, Explosives and Other Munitions of War.* 2d ed. Boston, 1882.

"Novoe ruzh'e i patron generala Berdana" (The new rifle and cartridge of General Berdan). *O.S.* 3 (1869): 32–40.

O. "Obzor deiatel'nosti voennogo ministerstva v poslednee piatiletie, finansovykh ego sredstv i nuzhd armii" (The work of the ministry of war for the past five years, its financial means, and the needs of the army). *V.S.* 10 (1865): 193–266.

"O merakh, priniatykh dlia vooruzheniia nashei armii skorostrel'nym oruzhiem" (Measures to arm our army with rifles). *O.S.* 1 (1869): 57–78.

"O revol'verakh" (On revolvers). *O.S.* 1 (1868): 24–43.

"O znachenii bystroi strel'by, o noveishikh obraztsakh magazinnykh ruzhei i priborakh dlia uskoreniia ruzheinoi strel'by" (The significance of rapid fire, the latest model magazine rifles, and the instruments to increase the rapidity of fire). *O.S.*, 6 installments (1879–1880).

Obozrenie Londonskoi Vsemirnoi vystavki po glavneishim otrasliam manufakturnoi promyshlennosti (A review of the major manufacturing sections of the London World Exposition). St. Petersburg, 1852.

"Obzor, kratkii, preobrazovanii po artillerii s 1856 po 1863 g." (A brief survey of the changes in artillery from 1856 to 1863). *O.S.* 1 (1863), supplement.

Ocherki preobrazovanii v artillerii v period upravleniia generala-adiutanta Barantsova, 1863–1877 gg. (The changes in artillery during the administration of Adjutant General Barantsov, 1863–1877). St. Petersburg, 1877.

Ogorodnikov, S. F., comp. *Istoricheskii obzor razvitiia i deiatel'nosti Morskogo ministerstva za sto let ego sushchestvovaniia* (A century of the naval ministry). Russian Naval Ministry. St. Petersburg, 1902.

"Opisanie poslednikh izmenenii v mekhanizme 4.2-lineinykh vintovok so skol'ziashchim zatvorom" (Recent changes in the mechanism of the .42-caliber bolt-action rifle). *O.S.* 3 (1877): 47.

"Opisanie razrabotok, upotrebliaiushchikhsia na Sestroretskom oruzheinom zavode dlia izgotovleniia zamochnykh chastei nashei shestilineinoi vintovki" (The procedures used at the Sestroretsk armory to make lock parts for the .60-caliber rifle). *O.S.* 1 (1869): 23–40.

Opisanie russkoi malokalibernoi vintovki vsekh obraztsov so skol'ziashchim zatvorom vtoroi sistemy Berdana (A description of all the models of the Berdan 2 Russian small-caliber bolt-action rifle). St. Petersburg, 1875.

"Opisanie 30-letnego iubileia Izhevskogo zavoda" (The thirtieth anniversary of the Izhevsk armory). *A. Zh.* (1857).

"Opyty nad malokalibernymi ruzh'iami Berdana, vtorogo obraztsa" (Tests on the Berdan 2 small-caliber rifle). *O.S.* 1 and 2 (1872): 26–48 and 88–111.

Orfeev, A. "Istoriia Sestroretskogo oruzheinogo zavoda" (History of the Sestroretsk armory). *O.S.*, 8 installments (1900–1904).

Ostroverkhov and Larionov. "Kurs o ruchnom ognestrel'nom oruzhii, sostavlennyi po lektsiiam, chitannym v strelkovoi ofitserskoi shkole v 1858 i 1859 gg." (A course on firearms based on lectures given at the officer's sharpshooting school in 1858 and 1859). *V.S.*, 4 installments (1859).

"Otchet nachal'nika artillerii Voennomu Ministru SShA za 1872 g." (A report of the commander of artillery to the U.S. minister of war for 1872). *A. Zh.*, 7 installments (1873–1875).

"Otchet za trekhletnee upravlenie artilleriei Ego Imperatorskim Vysochestvom Generaly-Fel'dtseikhmeisterom, s 25go ianvaria 1856 po 25 ianvaria 1859" (A three-year report on the administration of the artillery by the director-general of ordnance from 25 January 1856 to 25 January 1859). *A. Zh.* 1–6 (1860): 1–383.

P., L. "Kavaleriiskii pistolet polkovnika Lilienfel'da" (Colonel Lilienfel'd's cavalry pistol). *R.I.* 109 (19 May 1870): 4.

————. "Revol'ver, kak oruzhie dlia kavalerii" (The revolver as a cavalry arm).
    VS 9 (1866): 25–52.
Pekhotnyi revol'ver 3-go obraztsa sistemy Smita i Vesona (The Model 3 Smith
    and Wesson infantry revolver). St. Petersburg, 1876.
Pelenkin, A. "Fabrikatsiia tsel'no-tianutykh metallicheskikh patronov
    sistemy Berdana, shestilineinogo kalibra" (Fabrication of the Berdan
    .60-caliber drawn-brass metallic cartridge). *O.S.* 1 and 2 (1872): 26–48 and
    88–111.
"Perevooruzhenie nashei armii" (The rearmament of our army). *R.I* 260 (21
    November 1871): 1–3.
"Pistolet Smita i Vesona obraztsa 1871" (The Smith and Wesson 1871 model
    pistol). *O.S.* 4 (1872): 98–124.
Ploennis, Wilhelm. "Igolchatoe oruzhie: Materialy dlia kritiki oruzhiia, zaria-
    zhaiushchegosia s kazennoi chasti" (The needle gun: materials for a critical
    evaluation of breechloading firearms). *O.S.*, 7 installments (1867–1868).
    Translated from the German.
————. *Novye issledovaniia nad nareznym ognestrel'nym oruzhiem* (New re-
    search on rifles). St. Petersburg, 1863. Translated from the German.
"Po voprosu o vooruzhenii nashei armii skorostrel'nymi ruzh'iami" (The re-
    armament of our army with firearms). *VS.* 11 (1866): 33–43.
Popov, A. "Neobkhodimost' zameny revol'verov sistemy Smita i Vesona" (The
    need to replace the Smith and Wesson revolver). *O.S.* 4 (1891): 1–5.
Pototskii, N. P. *Sovremennoe ruchnoe oruzhie, ego svoistva, ustroistvo i
    upotreblenie; rukovodstvo, prisposoblennoe k programmam iunkerskikh
    uchilishch i teoreticheskogo kursa uchebnogo bataliona* (The characteris-
    tics, structure, and usage of modern firearms; a manual adapted to the
    programs of military academies and of theoretical courses of the training
    battalion). St. Petersburg, several editions.
————. *Turetskie ruzh'ia vo vremia 1877 g. Sravnenie turetskogo vooruzheniia
    s russkimi* (Turkish arms during the war of 1877. A comparison of Turkish
    and Russian arms). St. Petersburg, 1878. Reprint from *Zapiski R. T. O.*
    2 (1878): 118–44.
"Prodolzhenie opytov nad malokalibernym ruzh'iam Berdana vtorogo ob-
    raztsa" (A continuation of tests of the Berdan model 2 small-caliber rifle).
    O.S. 3 (1871): 27–56.
Report from the Select Committee on Small Arms (British Parliamentary
    Papers—Reports from Committees, vol. 12). London, 1854.
*The Rifle Conference of 1864: Report and Proceedings, with additional papers,
    drawings and index.* London, 1864.
Romanov, A. "Izhevskii oruzheinyi zavod: Mediko-topograficheskii ocherk"
    (The Izhevsk small arms factory: a medical and topographical study). *Sbor-
    nik sochinenii po sudebnoi meditsine* 3 (1875): 1–37.
Rüstow, Caesar. *Voennoe ruchnoe ognestrel'noe oruzhie* (Military firearms).
    St. Petersburg, 1861. Translated from the German.
"Ruzh'ie sistemy Pibodi, zariazhaiushcheesia s kazni metallicheskimi patro-

nami" (The Peabody breechloading rifle with metallic cartridges). *O.S.* 4 (1866): 65–77.

Sartisson, F. *Beiträge zur Geschichte und Statistik des russischen Bergbau- und Huttenwessens* (Materials on the history and statistics of the Russian mining and smelting industries). Heidelberg, 1900.

Sbornik noveishikh svedenii o ruchnom ognestrel'nom oruzhii dlia gg. pekhotnykh i kavaleriiskikh ofitserov russkoi armii (A collection of the latest information on firearms for infantry and cavalry officers of the Russian army). St. Petersburg, 1857.

Sbornik svedenii ofitserov Nikolaevskoi Akademii General'nogo Shtaba (A collection of the writings of the officers of the Nikolaev Academy general staff). 2 vols. St. Petersburg, 1862, 1863.

Schoen, Joseph. *Kratkoe obozrenie noveishikh sistem oruzhiia i primeneniia ikh k vooruzheniiu pekhoty v razlichnykh evropeiskikh gosudarstvakh* (Rifled infantry arms; a brief description of the modern systems of small arms as adopted in the various European armies). St. Petersburg, 1858. Translated from the German.

Shkliarevich, V. "Ob izuchenii svoistv ognestrel'nogo oruzhiia" (The study of the characteristics of firearms). *O.S.* 7 (1869): 209–54.

———. "Sanktpeterburgskaia masterskaia metallicheskikh patronov" (The St. Petersburg metallic cartridge shop). *R.I.* 23 (22 February 1869): 3–4.

Subbotkin, M. "Ob Izhevskom oruzheinom zavode" (The Izhevsk armory). *O.S.* 2 (1863): 150–78.

Svetlitskii, N. "Vospitanie voisk: Neskol'ko slov o vliianii nareznogo oruzhiia na boevoe znachenie kavalerii" (The training of soldiers: a comment on the influence of rifles on the military significance of the cavalry). *O.S.* 1 (1863): 1–44.

Terent'ev, Mikhail. "Kavaleriiskie voprosy" (Cavalry questions). *V.S.*, 4 installments (1865).

———. "Vzgliad na istoriiu i sovremennoe sostoianie povtoritel'nogo oruzhiia, ili revol'verov" (A look at the history and modern state of the repeating firearm, or revolver). *V.S.* 12 (1860): 215–68.

———. "Zametki o vooruzhenii kavalerii" (Notes on the armament of the cavalry). *V.S.* 2 (1860): 405–14.

Tenner, E. K. "Patronnoe proizvodstvo za granitsei i u nas" (Cartridge manufacture at home and abroad). *O.S.*, 5 installments (1880–1881).

"Tul'skii oruzheinyi zavod" (The Tula small arms factory). *Otechestvennye zapiski* 1–2 (1818–1819): 95–122.

Turner, N. O. S., F. G. E. Warren, and J. P. Nolan. *Tour of Artillery Officers in Russia*. British War Office. London, 1867.

U.S. Bureau of Foreign and Domestic Commerce. *Special Consular Reports*. Nos. 1–86. Washington, D.C., 1890–1923.

U.S. Census Office. "Report on Manufactures of Interchangeable Mechanism." *Tenth Census*, vol. 11: *Manufactures*. Washington, D.C., 1883.

U.S. Congress. Senate. *The Armies of Europe. Report of General George B.*

McClellan. 35th Cong., special session. Exec. Doc. no. 1. Philadelphia, 1861.

―――. Senate. *Military Commission to Europe in 1855 and 1856. Report of Major Alfred Mordecai of the Ordnance Department.* 36th Cong., 1st sess. Exec. Doc. no. 60. Washington, D.C., 1860.

―――. Senate. *Report on the Art of War in Europe in 1854, 1855, and 1856 by Major Richard Delafield, Corps of Engineers.* 36th Cong., 1st. sess. Exec. Doc. no. 59. Washington, D.C., 1860.

Voennoe Ministerstvo (Ministry of War). *Ezhegodnik russkoi armii* (The Russian army annual). St. Petersburg, 1868–1881.

―――. *Peremeny v obmundirovanii i vooruzhenii voisk, 1855–1881* (Changes in the uniforms and weapons of the soldiers, 1855–1881). 25 vols. St. Petersburg, 1857–1881.

―――. *Sbornik materialov po russko-turetskoi voine 1877–1878 na Balkanskom poluostrove* (Collection of materials on the Russo-Turkish War of 1877–1878 on the Balkan peninsula). 97 vols. St. Petersburg, 1909–11.

―――. *Voenno-uchenyi arkhiv: Materialy* (The military and scientific archive: materials). St. Petersburg, 1871.

―――. Voenno-istoricheskaia komissiia glavnogo shtaba. *Opisanie russko-turetskoi voiny 1877–1878 gg. na Balkanskom poluostrov* (Description of the Russo-Turkish War of 1877–1878 on the Balkan Peninsula). St. Petersburg, 1901.

"Vooruzhenie nashei kavalerii" (The weapons of our cavalry). *O.S.* 2 (1871): 141–46.

Vorob'ev, S. *Novoe ruchnoe ognestrel'noe oruzhie evropeiskikh armii* (The new firearms of the European armies). St. Petersburg, 1864.

Vorob'ev, V. P. V. "Kakoe oruzhie nam nuzhno" (The weapon Russia needs). *O.S.* 4 (1867): 111–22.

―――. "Revol'ver, kak oruzhie stroevogo ofitsera" (The revolver as a line officer's weapon). *O.S.* 1 (1866): 1–31.

―――. "Zametki o revol'verakh" (Notes on revolvers). *O.S.* 1 (1869): 1–22.

Zagoskin, V. N. "O tekhnicheskikh usloviiakh deshevogo proizvodstva metallicheskikh patronnykh gil'z" (The technical conditions for the inexpensive manufacture of metallic cartridge cases). *O.S.* 3 (1875): 28–35.

―――. "O vvedenii v Rossii metallicheskogo patrona" (The adoption of metallic cartridges in Russia). *Zapiski R. T. O.* 3 (1871): 157–83; and 5 (1875): 189–200.

Zakharov, M. A. "O stoimosti 4.2 lineinykh gil'z v zavisimosti ot nekotorykh uslovii ikh fabrikatsii" (The cost of .42-caliber cartridge cases as a function of certain conditions of their fabrication). *Zapiski R. T. O.* 6 (1886): 263–91.

Zybin, S. A. *Istoriia Imp. Tul'skogo Oruzheinogo Zavoda* (History of the Imperial Tula armory). Moscow, 1912. Reprint from *O.S.,* 12 installments (1897–1909).

―――. "Perevooruzhenie ruzh'iami umen'shennogo kalibra" (Rearmament with small-caliber firearms). *O.S.* 3 (1894): 43–101.

Zybin, S., M. Nekliudov, and M. Levitskii. *Oruzheinye zavody: Tul'skii,*

Sestroretskii, Izhevskii (The Tula, Sestroretsk, and Izhevsk armories). Kronstadt, 1898.

Secondary Sources

Adamov, E. A. "Russia and the U. S. at the Time of the Civil War." *Journal of Modern History* 11 (1930): 586–611.

Afremov, Ivan F. *Istoricheskoe obozrenie tul'skoi gubernii* (A historical review of the Tula province). Moscow, 1850.

Akademiia nauk SSSR. Arkheograficheskaia komissiia (USSR Academy of Sciences. Archæographic Commission). *Krepostnaia manufaktura v Rossii. Tul'skie i Kashirskie zheleznye zavody* (Serf manufactures in Russia. The Tula and Kashira iron factories). Leningrad, 1930.

———. Institut estestvozananii i tekhniki (Institute of Natural Sciences and Engineering). *Istorii tekniki* (History of engineering). Moscow, 1962.

———. Institut istorii (Institute of History). *Ocherki istorii Leningrada* (Studies of the history of Leningrad). 5 vols. Moscow, 1955–1970.

Alabin, P. *Chetyre voiny: Pokhodnye zapiski v 1849, 1853, 1854–56, 1877–78 godakh* (Four wars: diaries from the campaigns of 1849, 1853, 1854–56, 1877–78). Samara, 1888.

Aleksandrov, A. A. *Izhevskii zavod: Nauchno-populiarnyi ocherk istorii zavoda, 1760–1917* (The Izhevsk armory: a popular history of the factory, 1760–1917). *Istoriia fabrik i zavodov* series, no. 4. Izhevsk, 1957.

———. "Voprosy voennoi promyshlennosti Rossii vtoroi poloviny XIX v. v sovetskoi istoricheskoi literature" (Problems regarding Russia's military industry during the second half of the 19th century in the Soviet historical literature). In *Istoricheskie nauki na Urale za 50 let, 1917–1967* (The historical sciences in the Urals from 1917 to 1967). Vol. 1. Sverdlovsk, 1967. Pp. 75–81.

Amann, R., J. M. Cooper, and R. W. Davies, eds. *The Technological Level of Soviet Industry.* New Haven, 1977.

"Amerikanskaia artilleriia" (American artillery). *Inzhenernyi zhurnal* 4 (1864): 600–23.

Ames, Edward, and Nathan Rosenberg. "The Enfield Arsenal in Theory and History." In S. B. Saul, ed. *Technological Change: U.S. and Great Britain in the 19th Century.* London, 1970. Pp. 99–119.

Ashurkov, V. N. "Arendo-kommercheskoe upravlenie russkimi oruzheinymi zavodami" (The commercial lease management of Russian arms factories). *Uchenye zapiski T.G.P.I.* 8 (1958): 27–48.

———. "Golos tul'skikh oruzheinikov" (The voice of the Tula armorers). *Katorga i ssylka* 2/99 (1933): 141–45.

———. "Izuchenie istorii Izhevskogo zavoda" (Histories of the Izhevsk armory). *Ural i problemy regional'noi istoriografii: Period kapitalizma* (The Urals and the problems of regional historiography: the period of capitalism). Sverdlovsk, 1986. Pp. 33–36.

———. "K istorii promyshlennogo perevorota v Rossii: Rekonstruktsiia gosudarstvennykh oruzheinykh zavodov vo vtoroi polovine XIX v." (The industrial revolution in Russia: the reconstruction of the government armories in the second half of the 19th century). *Uchenye zapiski T.G.P.I.* (1967): 104–26.

———. "K voprosu o putiakh razvitiia russkikh oruzheinykh zavodov v 60-kh godakh XIX v." (The development of Russian arms factories in the 1860s). *Uchenye zapiski T.G.P.I.* 4 (1953).

———. *Kuznitsa oruzhiia: Ocherki po istorii Tul'skogo oruzheinogo zavoda* (The weapons makers: a history of the Tula arms factory). *Istoriia fabrik i zavodov* series, no. 21. Tula, 1947.

———. "Russkie oruzheinye zavody v 90 gg. XIX veka" (Russian small arms factories in the 1890s). *Uchenye zapiski T.G.P.I.* 2 (1969): 114–35.

———. "Russkie oruzheinye zavody v 40–50kh gg. XIX v." (Russian small arms factories in the 1840s and 1850s). *Voprosy voennoi istorii Rossii* (Problems in Russian military history). Moscow, 1969. Pp. 204–15.

———. *Tul'skii muzei oruzhiia: Putevoditel'* (A guidebook to the Tula arms museum). Tula, 1972.

———. "Vvedenie avtomaticheskogo oruzhiia v russkoi armii: Voennoe vedomstvo i kontsern 'Vikkers-Maksim'" (The adoption of the automatic rifle in the Russian army: the war office and the Vickers-Maxim company). In *Iz istorii Tul'skogo kraia* (The history of the Tula region). Tula, 1972.

Askew, William C. "Russian Military Strength on the Eve of the Franco-Prussian War." *Slavonic and East European Review* 30, 74 (December 1951): 185–205.

Avdeev, V. A., P. A. Zhilin, et al., eds. *Russkaia voennaia mysl' konets XIX-nachale XX v.* (Russian military thought in the late 19th and early 20th centuries). Moscow, 1982.

Bakulev, G., and D. Solomentsev, *Promyshlennost' Tul'skogo ekono-micheskogo raiona* (The industry of the Tula economic region). Tula, 1960.

Battison, Edwin. "Searches for Better Manufacturing Methods." *Tools and Technology* 3, 4 (Winter 1979): 13–18.

Beaver, Daniel R. "Cultural Change, Technological Development and the Conduct of War in the Seventeenth Century." In Russell F. Weigley, ed. *New Dimensions in Military History*. San Rafael, Calif., 1975.

Becker, William H. "American Manufactures and Foreign Markets, 1870–1900." *Business History Review* 47, 4 (Winter 1973): 466–81.

Beskrovnyi, L. G. *Armiia i flot Rossii v nachale XX v: Ocherki voenno-ekono-micheskogo potentsiala* (The Russian army and navy at the beginning of the 20th century: studies in military and economic potential). Moscow, 1986.

———. *Ocherki po istochnikovedeniiu voennoi istorii Rossii* (A source study for Russian military history). Moscow, 1957.

———. *Russkaia armiia i flot v XIX v.* (The Russian army and navy in the 19th century). Moscow 1973.

Bestuzhev-Riumin, V. "Ruchnoe oruzhie na parizhskoi mezhdunarodnoi vystavke" (Small arms at the Paris International Exposition). *O.S.* 4 (1867): 35–68.

Beyrau, Dietrich. *Militär und Gesellschaft im vorrevolutionären Russland.* Cologne and Vienna, 1984.

Bishop, John L. *A History of American Manufacturers from 1608 to 1860.* 3 vols. Philadelphia, 1866.

Blackmore, Howard L. "Colt's London Armoury." In S. B. Saul, ed. *Technological Change: U.S. and Great Britain in the 19th Century.* London, 1970. Pp. 171–95.

Blackwell, William L. *The Beginnings of Russian Industrialization, 1800–1860.* Princeton, 1968.

Bol'dt, K. "Rukovodstvo k izucheniiu okhotnich'ego oruzhiia" (A manual for hunting guns). *O.S.* 2–4 (1863).

Borisov, V. M. "Istoriia razvitiia kustarnykh promyslov v g. Tule i Tul'skom uezde i mery k dal'neishemu razvitiiu promyslov" (The history of the handicraft industries of Tula and Tula County and measures for their future development). *Trudy Komissii po issledovaniiu kustarnoi promyshlennosti Rossii* (Proceedings of the Commission for the Study of the Handicraft Industries of Russia). Vol. 9. St. Petersburg, 1883.

Brandenburg, N. E. *Istoricheskii katalog Sankt-Peterburgskogo artilleriiskogo muzeia* (Historical catalogue of the St. Petersburg artillery museum). St. Petersburg, 1877.

———. *500-letie russkoi artillerii, 1389–1889 gg.* (The 500th anniversary of Russian artillery, 1389–1889). St. Petersburg, 1889.

Bremner, Robert. *Excursions in the Interior of Russia.* 2 vols. London, 1839.

Britkin, A. S. *The Craftsmen of Tula: Pioneer Builders of Water-Driven Machinery.* Jerusalem, 1967.

Brown, Lawrence. *Innovation Diffusion: A New Perspective.* London, 1981.

Brown, Shannon R. "The Ewo Filature: A Study in Transfer of Technology to China in the 19th century." *T.C.* 20, 3 (July 1979): 550–68.

———. "The Transfer of Technology to China in the Nineteenth Century." *Journal of Economic History* 39 (1979): 181–97.

Bubenkov, K. D. "Perevooruzhenie russkoi armii v 1860–1870 gg." (The rearmament of the Russian army, 1860–1870). Candidate's diss. Leningrad, 1952.

Budaevskii, S. *Kurs artillerii* (Textbook of artillery). 8th ed. 5 vols. St. Petersburg, 1912.

Bumagin, A. A., ed. *Voenno-istoricheskii muzei artillerii i inzhenernykh voisk: Kratkii putevoditel'* (A guidebook to the military-historical museum of artillery and the corps of engineers). 4th ed. Leningrad, 1964.

Buttrick, John. "The Inside Contract System." *Journal of Economic History* 12, 3 (Summer 1952): 205–21.

[Calthorpe, S. J. G.] *Letters from Headquarters, or, the Realities of the War in Crimea.* 2 vols. London, 1856.

Carden, Godfrey L., comp. U.S. Bureau of Foreign and Domestic Commerce. Special Agents Series. *Machine-tool Trade in Austria-Hungary, Denmark, Russia and the Netherlands.* Washington, D.C., 1910.

Carey, Arthur Merwyn. *American Firearms Makers.* New York, 1953.

Carstensen, Fred V. *American Enterprise in Foreign Markets: Studies of Singer and International Harvester in Imperial Russia.* Chapel Hill, N.C., 1984.

Carstensen, Fred, and Richard Hume Working. "International Harvester in Russia: A Washington-St. Petersburg Connection?" *Business History Review* 57, 3 (Autumn 1983): 347–66.

Carver, J. Scott. "*Tekhnologicheskii zhurnal*: An Early Russian Techno-economic Periodical." *T.C.* vol. 18 (October 1977): 622–43.

Cesari, Gene Silvero. "American Arms-Making Machine Tool Development, 1798–1855." Ph.D. diss. University of Pennsylvania, 1970.

Chebyshev, V. "Glavneishie voprosy po ruzheinoi chasti" (The most important problems in small arms) *O.S.* 3 (1882): 35–64; 2 (1883): 1–27; 2 and 3 (1885): 1–25 and 1–25; 2 and 3 (1886): 1–19 and 1–46.

———. "Publichnye lektsii, chitannye pri gvardeiskoi artillerii" (Public lectures delivered at the guard artillery). *A. Zh.* 3 and 4 (1861): 157–212 and 213–39.

Chinn, G. *The Machine Gun.* Washington, 1955.

Chizhikov, N. A. "Noveishie usovershenstvovaniia v okhotnich'em oruzhii" (The latest improvements in hunting arms). *Zapiski R. T. O.* 2 (1886): 62–110.

Clarke, F. C. H. "Recent Reforms in the Russian Army." *Journal of the Royal United Service Institution* 20, 86 (1876): 373–88.

Cleinow, George. *Beiträge zur lage der hausindustrie in Tula* (Materials on the state of the domestic industries in Tula). Leipzig, 1904.

Coleman, Marion Moore. "Eugene Schulyer: Diplomat Extraordinary from the U.S. to Russia, 1867–1876." *Russian Review* 7 (1947): 33–48.

Crisp, Olga. "Labor and Industrialization in Russia." *Cambridge Economic History of Europe*, vol. 7, pt. 2. Cambridge, 1978. Pp. 308–415.

Crouzet, François. "Recherches sur la production d'armements en France, 1815–1913." In *Conjuncture économique, structures sociales: Hommage à Ernest Labrousse.* Paris, 1974.

Curtin, Jeremiah. *Memoirs.* Madison, Wis., 1940.

Curtiss, John S. *The Army of Nicholas I.* Durham, N.C., 1965.

D–r, R. "Ruchnoe ognestrel'noe oruzhie so vremeni voiny 1870–71 gg." (Firearms since the war of 1870–71). *O.S.* 4 (1878): 87–143; 3 and 4 (1879): 78–92 and 1–35; 1 and 2 (1882): 41–61 and 59–86.

Daniels, George. "The Big Questions in the History of American Technology." *T.C.* 11 (1970): 1–35.

Daumas, Maurice, ed. *A History of Technology and Invention: Progress through the Ages.* Trans. Eileen B. Hennessy. 4 vols. New York, 1979.

Denisova, M. M. *Russkoe oruzhie: Kratkii opredelitel' russkogo boevogo oruzhiia XI-XIX vv.* (A short dictionary of Russian weapons from the 11th to the 19th centuries). Moscow, 1953.

Deyrup, Felicia Johnson. *Arms Makers of the Connecticut Valley.* Smith College Studies in History. Northampton, Mass., 1948.

Drake, Mervin. "Breechloading rifles and the governments of France, Prussia, and England." *Journal of the Royal United Services Institution* 15, 64 (1871): 438.

Dupree, A. Hunter. "Does the History of Technology Exist?" *Journal of Inter-disciplinary History* 11, 4 (Spring 1981): 685–94.

Dupuy, R. Ernest, and Trevor Dupuy. *Encyclopedia of Military History.* London, 1977.

Edwards, W. B. *The Story of Colt's Revolver: A Biography of Samuel Colt.* Harrisburg, 1953.

Ellis, John. *The Social History of the Machine Gun.* New York, 1981.

Ermoshin, I. P., ed. *Artilleriiskii istoricheskii muzei: Putevoditel' po istoricheskomu arkhivu muzeia* (The museum of artillery history: guidebook to the museum's historical archives). Leningrad, 1957.

Esper, Thomas. "Industrial Serfdom and Metallurgical Technology in 19th-Century Russia." *T.C.* 23 (October 1982): 583–608.

Die Europäischen Heere, ihre Organisierung und Bewaffnung (The organization and armament of the European armies). Hildburghausen, 1870.

Ezell, Edward C. *The AK 47 Story: Evolution of the Kalashnikov Weapons.* Harrisburg, Pa., 1986.

———. *Handguns of the World: Military Revolvers and Self-Loaders from 1870 to 1945.* Harrisburg, Pa., 1981.

Fadeev, Rotislav A. *Vooruzhennye sily Rossii* (The armed forces of Russia). Moscow, 1868.

Falls, Cyril. *A Hundred Years of Warfare.* London, 1953.

Fedorov, A. V. *Russkaia armiia v 50–70kh gg. XIX v.* (The Russian army, 1850–1870). Leningrad, 1959.

Fedorov, V. G. *Evoliutsiia strelkogo oruzhiia* (The evolution of the rifle). Moscow, 1938.

———. "Ruchnoe ognestrel'noe oruzhie russkoi armii za XIX v." (Firearms of the Russian army in the 19th century). *O.S.*, 9 installments (1901–1903).

———. *Vooruzhenie russkoi armii XIX v.* (The arms of the Russian army in the 19th century). St. Petersburg, 1911.

———. *Vooruzhenie russkoi armii v Krymskuiu kampaniiu* (The arms of the Russian army in the Crimean campaign). St. Petersburg, 1904.

———. *Istoriia vintovki* (The history of the rifle). Moscow, 1940.

"Firearms." *Colburn's United Service Magazine* 370 (September 1859): 51.

"The Firearms Manufacture." American Industries, no. 75. *Scientific American,* 3 September 1881, p. 148.

Floud, Roderick. *The British Machine Tool Industry, 1850–1914.* London, 1976.

———. "Changes in the Productivity of Labour in the British Machine Tool Industry, 1856–1900." In Donald N. McCloskey, ed. *Essays on a Mature Economy: Britain after 1840.* Princeton, 1971.

Fon-der-Khoven, A. "Zametka o Providenskom oruzheinom zavode i o ruzh'iakh Pibodi-Martini, Turetskogo obraztsa" (A note on the Providence arms factory and on the Turkish model Peabody-Martini rifle). *O.S.* 4 (1877): 16–30.

Fries, Russell I. "British Response to the American System: The Case of the Small Arms Industry after 1850." *T.C.* 16, 3 (July 1975): 377–403.

————. "A Comparative Study of the British and American Arms Industries, 1790–1890." Ph.D. diss. Johns Hopkins University, 1972.

Fuller, J. F. C. Armament and History. New York, 1945.

————. A Military History of the Western World. 3 vols. New York, 1956.

Gaier-Lhoest, Claude. Four centuries of Liège Gunmaking. Trans. F. J. Norris. London, 1976.

Gatrell, Peter. The Tsarist Economy, 1850–1917. New York, 1986.

Gentry, Curt. John M. Browning, American Gunmaker. New York, 1964.

Geyer, Dietrich. Russian Imperialism. Leamington Spa, 1986.

Gilbert, K. R. "The Ames Recessing Machine: A Survivor of the Original Enfield Rifle Machinery." T.C. 9, 2 (Spring 1963): 207–11.

Gil'en, "Istoriia ruchnogo ognestrel'nogo oruzhiia" (A history of firearms). A. Zh. 2 (1858): 247–84.

Gluckman, Arcadi, and L. D. Satterlee. American Gun Makers. Harrisburg, 1953.

Gnatovskii, N. I., and P. A. Shorin. Istoriia razvitiia otechestvnnogo strelkovogo oruzhiia (A history of Russian rifles). Moscow, 1959.

Golder, Frank A. "The American Civil War through the Eyes of a Russian Diplomat." American Historical Review 26 (1921): 454–63.

————. Guide to Materials for American History in Russian Archives. 2 vols. Washington, 1917, 1937.

————. "Russian-American Relations during the Crimean War." American Historical Review 31 (April 1926): 462–76.

Goldman, Marshall. "The Relocation and Growth of the Pre-Revolutionary Russian Ferrous Industry." Explorations in Entrepreneurial History 9, 1 (1956): 19–36.

Goldstein, E. R. "Military Aspects of Russian Industrialization: The Defense Industries, 1890–1917." Ph.D. diss. Case Western Reserve University, 1971.

————. "Vickers Ltd. and the Tsarist Regime." Slavonic and East European Review 58, 4 (October 1980): 561–71.

Gorbov, M. I. Izhevskie oruzheiniki (The Izhevsk armorers). Istoriia fabrik i zavodov series, no. 100. Izhevsk, 1963.

Gorlov, Aleksandr. "Voennye ocherki SShA" (Military sketches of the USA). A. Zh. 7, 8, 10 (1866): 223–84, 441–66, 597–601.

Graham, Sir Lumley. "The Russian Army in 1882." Journal of the Royal United Services Institution, 4 installments (1883–1884).

Granick, David. Soviet Metal-Fabricating and Economic Development: Practice versus Policy. Madison, Wis., 1967.

Grant, Ellsworth S. "Gunmaker to the World." American Heritage 19, 4 (June 1968): 1–11, 86–91.

Greene, Major-General Francis V. The Russian Army and Its Campaigns in Turkey, 1877–78. 2 vols. London, 1879.

————. Sketches of Army Life in Russia. London, 1881.

Greifenfel'z, V. Giber von. Nakanune perevooruzheniia (On the eve of the rearmament). Sestroretsk, 1910.

Grierson, Captain J. M. The Armed Strength of Russia. London, 1882.

Grishinskii, A. S., ed. *Istoriia russkoi armii i flota* (A history of the Russian army and navy). 15 vols. Moscow, 1911.

Grodnitskii, A. N. "Fabrikatsiia bessemerovskoi stali dlia izgotovleniia voennogo oruzhiia" (The fabrication of Bessemer steel for military small arms). *Zapiski R. T. O.* 2 (1875): 25–52.

Guroff, Gregory, and Fred V. Carstensen. *Entrepreneurship in Imperial Russia and the Soviet Union.* Princeton, 1983.

Haas, H. van der. *The Enterprise in Transition: An Analysis of European and American Practice.* London, 1967.

Habakkuk, H. J. *American and British Technology in the Nineteenth Century.* Cambridge, 1962.

Hacker, Barton C. "The Weapons of the West: Military Technology and Modernization in 19th-century China and Japan." *T.C.* 18 (January 1977): 43–55.

Hamley, E. B. "The Armies of Russia and Austria." *Nineteenth Century* 3 (May 1878): 844–62.

Hanson, Philip. *Trade and Technology in Soviet-Western Relations.* New York, 1981.

Hart, B. H. Liddell. "Armed Forces and the Art of War: Armies." In *The New Cambridge Modern History,* vol. 10. Cambridge, 1960. Pp. 302–20.

——. *The Revolution in Warfare.* New Haven, 1947.

Hatch, Alden. *Remington Arms in American History.* New York, 1956.

Hogan, Heather. "Industrial Rationalization and the Roots of Labor Militance in the St. Petersburg Metalworking Industry, 1901–14." *Russian Review* 42, 2 (April 1983): 163–90.

——. "Labor and Management in Conflict: The St. Petersburg Metalworking Industry, 1900–1914." Ph.D. diss. University of Michigan, 1981.

Hounshell, David A. *From the American System to Mass Production, 1800–1932: The Development of Manufacturing Technology in the United States.* Studies in Industry and Society, no. 4. Baltimore, 1984.

Howard, Michael. *The Franco-Prussian War.* New York, 1961.

Howard, Robert A. "Interchangeable Parts Reexamined: The Private Sector of the American Arms Industry on the Eve of the Civil War." *T.C.* 19, 4 (October 1978): 633–49.

Hughes, Thomas P. "Emerging Themes in the History of Technology." *T.C.* 20 (October 1979): 697–711.

——. "The Order of the Technological World." *History of Technology* 5 (1980): 1–16.

Iakovlev, M. N. "*Voennyi sbornik* v period voennykh reform, 1860–70gg." (*Voennyi sbornik* during the miltary reforms, 1860–1870). *V.L.U.* 3 (1983): 32–36.

Ianevich-Ianevskii. A. K. "O sovremennom nareznom okhotnich'em oruzhii v Severnoi Amerike" (Modern hunting rifles in North America). *O.S.* 3 (1894): 1–41; 4 (1894): 1–54.

Iatsunskii, V. K. "Krupnaia promyshlennost' Rossii v 1790–1860 gg." (Heavy industry in Russia, 1790–1860). In M. K. Rozhkova, ed. *Ocherki ekonomicheskoi istorii Rossii pervoi poloviny XIX v.* (Studies in Russian Economic History during the first half of the 19th century). Moscow, 1959.

"Igol'chatye ruzh'ia v zapadnoi Evrope i u nas" (The needle gun in Western Europe and in Russia). *R.I.* 254 (6/18 October 1866): 3.

Istoriia rabochikh Leningrada, 1703–1965 (A history of the workers of Leningrad, 1703–1965). 2 vols. Leningrad, 1972.

Istoriia Tul'skogo oruzheinogo zavoda, 1712–1972 (History of the Tula arms factory). Moscow, 1973.

Ivanov, P. A. *Obozrenie sostava i ustroistva reguliarnoi kavalerii* (The composition and organization of the regular cavalry). St. Petersburg, 1864.

Izhevsk, 1760–1985: Dokumenty i materialy (Izhevsk, 1760–1985: documents and materials). Izhevsk, 1984.

Jinks, Roy G. *History of Smith and Wesson.* North Hollywood, Calif., 1977.

———. *Smith and Wesson, 1857–1945.* New York, 1975.

Jones, Archer. *The Art of War in the Western World.* Urbana and Chicago, 1987.

Jones, David. "Imperial Russia's Forces at War." In Allan R. Millet and Williamson Murray, eds. *Military Effectiveness,* vol. 1: *The First World War.* Boston, 1988. Pp. 249–328.

Jones, G., and C. Trebilcock. "Russian Industry and British Business, 1910–1930: Oil and Armaments." *Journal of Economic History* 11 (1982): 61–103.

Kahan, Arcadius. "Entrepreneurship in the Early Development of Iron Manufacturing in Russia." *Economic Development and Cultural Change* 10, 4 (1962): 395–422.

Kaser, M. C. "Russian Entrepreneurship." *The Cambridge Economic History of Europe,* vol. 7, pt. 2. Cambridge, 1978. Pp. 416–93.

Kashin, V. N. "Tul'skaia oruzheinaia sloboda v XVII v." (The Tula armorers' settlement in the 17th century). *Problemy istorii dokapitalisticheskogo obshchestva* (Problems in the history of pre-capitalist society) 5, 1–2 (1935): 11–141; 5–6 (1935): 76–99.

Katz, James Everett, ed. *Arms Production in Developing Countries: An Analysis of Decision Making.* Lexington, Mass., and Toronto, 1984.

Keegan, John. *The Face of Battle.* London, 1976.

Kennett, Lee, and James Laverne Anderson. *The Gun in America.* Westport, Conn., 1975.

Kersnovskii, A. *Istoriia russkoi armii* (A history of the Russian army). 2 vols. Belgrade, 1934–1938.

Kihn, Phyllis. "Colt in Hartford." *Connecticut Historical Society Bulletin* 24, 3 (July 1959): 74–87.

Kipp, Jacob W. "Consequences of Defeat: Modernizing the Russian Navy, 1856–63." *Jahrbücher für Geschichte Osteuropas* 20 (1972): 210–25.

———. "A Few Comments regarding Historical Sources on the Tsarist Navy during the Reigns of Nicholas I and Alexander II, 1825–81." *Military Affairs* 36 (December 1972): 127–30.

Kirchner, Walther. "The Industrialization of Russia and the Siemens Firm,

1853–1890." *Jahrbücher für Geschichte Osteuropas* 22, 3 (1974): 321–57.

———. "One Hundred Years of Krupp and Russia, 1818–1918." *Vierteljahrschrift für Sozial- und Wirtschaftsgeschichte* 69 (1982): 75–108.

———. "Russian Entrepreneurship and the Russification of Foreign Enterprise." *Zeitschrift für Unternehmensgeschichte* 26 (1981): 79–103.

———. *Studies in Russian-American Commerce, 1820–60.* Studien zur Geschichte Osteuropas, vol. 19. Leiden, 1975.

———. "Western Businessmen in Russia: Practices and Problems." *Business History Review* 38 (1964): 315–27.

Kitchener, H. E. C. "Revolvers and their use." *Journal of the Royal United Services Institution* 30, 136 (1886): 951–95.

Kohl, Johann. *Russia and the Russians in 1842.* London, 1842.

Kolchak, Vasilii I. *Istoriia Obukhovskogo staleliteinogo zavoda v sviazi s progressom artilleriiskoi tekhniki* (A history of the Ohukhov steel mill and the progress in artillery technology). St. Petersburg, 1903.

———. "Sovremennoe stal'noe delo na Obukhovskom zavode" (The Obukhov steel mill today). *M.S.* 8 and 9 (1875): 67–104, 1–30.

Kol'devin, N. *Bitva russkikh s bukhartsam v 1868 i geroicheskaia oborona Samarkanda* (The Russian-Bukhara Battle of 1868 and the heroic defense of Samarkand). St. Petersburg, 1873.

Kononova, N. N. "Rabochie oruzheinykh zavodov voennogo vedomstva v pervoi polovine XIX v." (Workers at the government armories during the first half of the 19th century). *Uchenye zapiski Leningradskogo gosudarstvennogo universiteta* 32, 270 (1959): 118–44.

Korolev, A. A. "Iz istorii russkoi voennoi promyshlennosti; vozniknovenie Tul'skogo chastnogo patronnogo zavoda" (From the history of Russian military industry; the founding of the Tula private cartridge factory). *Uchenye zapiski T.G.P.I.* 2 (Tula, 1969): 78–98.

Kovalevskii, E. *Voina s Turtsiei* (The Turkish War). St. Petersburg, 1871.

Kovalevskii, M. "Sovershenstvovaniia ruchnogo ognestrel'nogo oruzhiia v techenie XIX v." (Improvements in firearms during the 19th century). *O.S.* 2 (1903): 65–86.

Krylov, I. N. *Opisanie Imperatorskogo Tul'skogo oruzheinogo zavoda* (Description of the Imperial Tula small arms factory). Tula, 1903.

Kryzhanovskii, N. *Publichnye chteniia Generala Kryzhanovskogo, chitannye pri Gvardeiskoi Artillerii v 1858 g.* (The public lectures of General Kryzhanovskii at the guard artillery in 1858). St. Petersburg, 1858.

Kukushkin, V. N. *Sestroretskaia dinastiia: Ocherki o proshlom i nastoiashchem Sestroretskogo instrumental'nogo zavoda im. S. P. Voskova* (A Sestroretsk dynasty: studies of the S. P. Voskov Sestroretsk tool plant, past and present). Istoriia fabrik i zavodov series, no. 236. Leningrad, 1959.

Kuropatkin, A. "Artilleriiskie voprosy" (Artillery questions). *V.S.* 5 (1885): 58–116.

Lacombe, M. "Introduction du machinisme dans les fabrications d'armements en France au XIXe siècle" (The mechanization of French armament production in the 19th century). *Revue Internationale d'Histoire Militaire* 41 (1979): 37–48.

Landau, Sarah Bradford. "The Colt Industrial Empire in Hartford." *Antique* 3 (March 1976): 568–79.

Laskovskii, F. *Materialy dlia istorii inzhenernogo iskusstva v Rossii* (The history of the engineering arts in Russia). 3 vols. St. Petersburg, 1858–1865.

Latham, John. "Progress of Small Breechloading Arms." *Journal of the Royal United Services Institution* 19, 83 (1875): 631–53.

Layton, Edwin T., Jr., ed. *Technology and Social Change in America*. New York, 1973.

Leskov, Nikolai. "The Left-handed Craftsman: A Tale of the Cross-eyed, Left-handed Craftsman of Tula and the Steel Flea." In *Nikolai Leskov: Selected Tales*. Trans. David Magarshack. New York, 1961.

Levasheva, Z. M. *Bibliografiia russkoi voennoi bibliografii* (A bibliography of Russian military bibliography). Moscow, 1950.

Liapin, V. A. "Rabochie voennogo vedomstva vo vtoroi polovine XIX v." (Workers of the war office during the second half of the 19th century). In L. V. Olkhovaia, ed. *Genezis i razvitie kapitalisticheskikh otnoshenii na Urale* (The origin and development of capitalistic relations in the Urals). Sverdlovsk, 1980. Pp. 30–40.

Loubat, J. F. *Narrative of the Mission to Russia in 1866 of Hon. Gustavus Fox*. New York, 1873.

Lukin, P. "Kollektsiia ruchnogo ognestrel'nogo oruzhiia inostrannykh obraztsov, prinadlezhashchaia Oruzheinoi komissii" (The collection of foreign firearms in the possession of the small arms commission). *O.S.*, 3 installments (1861–1862).

Lundeberg, Philip K. *Samuel Colt's Submarine Battery: The Secret and the Enigma*. Smithsonian Studies in History and Technology, no. 29. Washington, 1974.

Luvaas, Jay. *The Military Legacy of the Civil War: The European Inheritance*. Chicago, 1959

MacGahan, Januarius. *Campaigning on the Oxus and the Fall of Khiva*. New York, 1874.

McKay, John P. "Foreign Businessmen, Tsarist Government and the Briansk Company." *Journal of Economic History* 2, 2 (Fall 1973): 273–93.

———. *Pioneers for Profit: Foreign Entrepreneurship and Russian Industrialization, 1885–1913*. Chicago, 1970.

McNeil, William. *The Pursuit of Power: Technology, Armed Force and Society since A.D. 1000*. Chicago, 1982.

Malkin, M. "K istorii russko-amerikanskikh otnoshenii vo vremia grazhdanskoi voiny v SShA" (Russian-American relations during the American Civil War). *Krasnyi arkhiv* 3/94 (1939): 97–153.

Markevich, V. E. *Ruchnoe ognestrel'noe oruzhie* (Firearms). Leningrad, 1937.

Marshall, A. C., and N. Newbould. *History of Firths*. Sheffield, 1924.

Matloff, Maurice. "The Nature and Scope of Military History." In Russell F. Weigley, ed. *New Dimensions in Military History*. San Rafael, Calif., 1975.

Maurice, Major F. B. *The Russo-Turkish War, 1877.* London, 1905.

Mavrodin, V. V. "K voprosu o perevooruzhenii russkoi armii v seredine XIX v." (The rearmament of the Russian army during the middle of the 19th century). *Problemy istorii feodal'noi Rossii.* Leningrad, 1971.

———. "Revol'very Tul'skikh oruzheinikov" (Revolvers of the Tula gunmakers). *Soobshcheniia Ermitazha* 40 (1975): 34–36.

———. "Strelkovoe oruzhie russkogo flota v XIX v." (Firearms of the Russian navy in the 19th century). *Morskoi sbornik* 9 (1978): 86–88.

Mavrodin, V. V., ed. *Rabochie oruzheinoi promyshlennosti v Rossii i russkie oruzheiniki v XIX-nachale XX v.* (Workers in the Russian arms industry and Russian gunmakers in the 19th and beginning of the 20th centuries). Leningrad, 1976.

Mavrodin, V. V., and Val. V. Mavrodin. *Iz istorii otechestvennogo oruzhiia: Russkaia vintovka* (The history of native firearms: the Russian rifle). Leningrad, 1981.

Mavrodin, V. V., and P. Sh. Sot. "Sovetskaia istoriografiia otechestvennogo strelkovogo oruzhiia XIX-nachale XX v." (Soviet historiography of the native firearms of 19th- and early 20th-centuries). *VLU* 14 (1976): 45–51.

Mavrodin, Val. V. "K voprosu o proizvodstve boepripasov k strelkovomu oruzhiiu v Rossii v 40–60 gg. XIX v." (Concerning the production of ammunition for rifles in the 1840s-1860s). *VLU* 2 (1974): 149–51.

———. "O nekotorykh oshibochnykh utverzhdeniiakh zarubezhnykh oruzhievedov" (Certain mistaken convictions of foreign arms experts). *VLU* 20 (1975): 137–38.

———."O priniatii na vooruzhenie russkoi armii 4.2-lineinoi vintovki: k stoletiiu russkoi berdanki" (The adoption of the .42-caliber rifle in the Russian army: the centenary of the Russian Berdan). *VLU* 4 (1969): 68–71.

———. "Ob odnoi oshibke v istoriografii Krymskoi voiny" (An error in the historiography of the Crimean War). *Istoriograficheskii sbornik* 6 (Saratov, 1977).

———. "Perevooruzhenie kazach'ikh voisk strelkovym oruzhiem vo vremia vtoroi polovine XIX v." (The rearmament of the Cossacks with rifles during the second half of the 19th century). *Izvestiia Severo-Kavkazskogo nauchnogo tsentra vysshei shkoly; Seriia obshchestvennykh nauk* 1 (1975): 77–82.

———. "Rabota F. Engel'sa 'Istoriia vintovki' i evoliutsiia sovremennogo strelkovogo oruzhiia" (Engels' 'History of the Rifle' and the evolution of modern firearms). *VLU* 20 (1970): 154–57.

Mayr, Otto, and Robert C. Post, eds. *Yankee Enterprise: The Rise of the American System of Manufacturing.* Washington, 1981.

Mendeleev, D. I., ed. *The Industries of Russia.* English edition ed. by John Martin Crawford. St. Petersburg, 1893.

Menning, Bruce. "Bayonets before Bullets: The Organization and Tactics of the Imperial Russian Army, 1861-1905." Unpublished Master of Military Art and Science Thesis. U.S. Army Command and General Staff College. Fort Leavenworth, Kansas, 1984.

Meshcheriakov, G. P. *Russkaia voennaia mysl' v XIX v.* (Russian military thought in the 19th century). Moscow, 1973.

Mezhenko, Iu. A. *Russkaia tekhnicheskia periodika, 1800–1916: Bibliograficheskii ukazatel'* (Russian technical periodicals, 1800–1916: bibliographical index). Moscow, 1955.

Mikheev, S. *Istoriia russkoi armii* (A history of the Russian army). 5 vols. Moscow, 1910–1911.

Miller, Forrestt. *Dmitrii Miliutin and the Reform Era in Russia.* Nashville, Tenn., 1968.

Mitchell, James L. *Colt, the Arms, the Man, the Company.* Harrisburg, 1959.

Moltke, Count Helmuth Karl Bernard von. *Field Marshal Count Moltke's Letters from Russia.* Trans. Robina Napier. London, 1878.

Mosse, W. E. *Alexander II and the Modernization of Russia.* New York, 1962.

Myshovskii, E. V. "O traktovke A. V. Fedorovym problem perevooruzheniia russkoi armii" (A. V. Fedorov's interpretation of the rearmament of Russia). *Sbornik issledovanii i materialov Artilleriiskogo istoricheskogo muzeia* 4 (Leningrad, 1959): 333–40.

———. "Russkoe avtomaticheskoe oruzhie: Raboty russkikh oruzheinikov, 1887–1917 gg." (The Russian automatic rifle: the work of Russian armorers, 1887–1917). *Sbornik issledovanii i materialov Artilleriiskogo istoricheskogo muzeia* 3 (Leningrad, 1958): 185–209.

Newman, J. R. *The Tools of War.* London, 1942.

Ocherki istorii tekhniki v Rossii, 1861–1917 (A history of technology in Russia, 1861–1917). Moscow, 1975.

Ocherki istorii Udmurtskoi ASSR (Studies in the history of the Udmurt ASSR). 2 vols. Izhevsk, 1958, 1962.

Otchet po venskoi vsemirnoi vystavke 1873 g. v voenno-tekhnicheskom otnoshenii (A report of the military aspects of the Vienna World Exposition of 1873). 2 vols. St. Petersburg, 1874 and 1879.

Paret, Peter, and Felix Gilbert, eds. *The Makers of Modern Strategy: From Machiavelli to the Nuclear Age.* Princeton, 1986.

Parker, J. E. S. *The Economics of Innovation: The National and Multinational Enterprise in Technological Change.* London, 1974.

Parson, John E. "Colt Brevete." *The American Rifleman* (June 1950): 22–36.

———. *The First Winchester.* New York, 1955.

———. *The New York Metropolitan Museum of Art Catalogue of a Loan Exhibition of Percussion Colt Revolvers and Conversions, 1836–1873.* New York, 1972.

———. *Smith and Wesson Revolvers.* New York, 1957.

Patin, K. *Spravochnik: Polnyi i podrobnyi ukazatel' vsekh deistvuiushchikh prikazov po voennomy vedomstvu, tsirkuliarov, predpisannii i otzyvov General'nogo shtaba i prochikh glavnykh upravlenii, i prikazov, prikazanii i tsirkuliarov po vsem voennym okrugam za 45 let, s 1859 po 1904 g.* (A reference guide: a full and detailed index to all the operational orders from the war office, circulars, directions, and evaluations of the general staff and other chief offices, and orders, directives, and circulars to all military districts for 45 years, from 1859 to 1904). 2 vols. Tambov, 1904.

Pavlov, M. A., ed. *Metallurgicheskie zavody na territorii SSSR s XVII do 1917 g.* (Metallurgical plants on the territory of the USSR from the 17th century to 1917). Moscow-Leningrad, 1937.

Peterson, Harold L. *The Remington Historical Treasury of American Guns.* New York, 1966.

Pintner, Walter McKenzie. "The Burden of Defense in Imperial Russia, 1725–1914." *Russian Review* 43, 3 (July 1984): 231–59.

———. *Russian Economic Policy under Nicholas I.* Ithaca, N.Y., 1967.

———. "The Russian Military, 1700–1917: Social and Economic Aspects." *Trends in History* 2, 2 (Winter 1981): 43–52.

———. "Russian Military Thought: The Western Model and the Shadow of Suvorov." In Peter Paret and Felix Gilbert, eds. *Makers of Modern Strategy.* Princeton, 1986. Pp. 354–75.

Platov and Kirpichev, L. *Istoricheskii ocherk obrazovaniia i razvitiia artilleriiskogo uchilishcha* (A history of the formation and development of the artillery school). St. Petersburg, 1870.

Ploennis, Wilhelm. "Vopros o ruchnom oruzhii v Germanii" (Firearms in Germany). O.S., 7 installments (1873–1875). Translated from the German.

Pollard, Sidney. *The Genesis of Modern Management.* Cambridge, Mass., 1965.

Portnov, M. K. "K istorii priniatiia na vooruzhenie russkoi armii 4.2 lineinoi vintovki obraztsa 1868 g." (The history of the adoption of the 1868 model .42-caliber rifle in the Russian army). *Ezhegodnik Gosudarstvennogo Istoricheskogo muzeia 1961 g.* Moscow, 1962. Pp. 63–70.

Pototskii, N. P., and P. Platonov. *Stoletie rossiiskoi artillerii, 1794–1894* (A century of Russian artillery, 1794–1894). St. Petersburg, 1894.

Pozdnev, A. *Tvortsy otechestvennogo oruzhiia* (Russian arms makers). Moscow, 1955.

Purves, J. G. "Nineteenth-century Russia and the Revolution in Military Technology." In J. G. Purves and D. A. Wests, eds. *War and Society in the 19th-century Russian Empire.* Toronto, 1972.

Reeve, Henry, ed. *St. Petersburg and London in the Years 1852–1864: Reminiscences of Count Charles Frederick Vitzthum von Eckstaedt, Late Saxon Minister at Court of Sir James.* 2 vols. London, 1887.

Rieber, Alfred J. "The Formation of La Grande Société des Chemins de Fer Russes.'" *Jahrbücher für Geschichte Osteuropas* 21, 3 (1973): 375–91.

Rieber, Alfred J., ed. *The Politics of Autocracy: The Letters of Alexander II to Prince A. I. Bariatinskii, 1857–1864.* Paris, 1966.

Rigley, John. "The Manufacturing of Small Arms." *Proceedings of the Institute of Civil Engineers,* vol. 111. London, 1893. Pp. 129–225.

Roads, C. H. *The British Soldier's Firearm, 1850–64.* London, 1964.

Robertson, James Road. *A Kentuckian at the Court of the Tsars: The Ministry of Cassius Clay to Russia, 1861–62 and 1863–69.* Berea, Ky., 1935.

Roe, Joseph Wickham. *English and American Tool-Builders*. New Haven, 1916.

Rogger, Hans. "Amerikanizm and the Economic Development of Russia." *Comparative Studies in Society and History* 23, 3 (July 1981): 382–420.

Rohan, Jack. *Yankee Arms Maker*. New York, 1935.

Rosa, Joseph. *Colonel Colt: London*. London, 1976.

Rosenberg, Nathan. *The American System of Manufactures*. Edinburgh, 1969.

———. *Perspectives on Technology*. Cambridge, 1976.

———. "Technological Interdependence in the American Economy." *T.C.* 20, 1 (January 1979): 25–50.

———. *Technology and American Economic Growth*. New York, 1972.

Rozenfel'd, Ia. S., and K. I. Klimenko. *Istoriia mashinostroeniia SSSR s pervoi poloviny XIX v. do nashikh dnei* (A history of machine building in the USSR from the first half of the 19th century to the present). Moscow, 1961.

Rozhkova, M. K., ed. *Ocherki ekonomicheskoi istorii Rossii pervoi poloviny XIX v.* (Russian economic history during the first half of the 19th century). Moscow, 1959.

"Samuel Colt and the Revolver." *Industrial America*. 1876.

Saul, S. B. "The Machine Tool Industry in Britain to 1914." *Business History* 10, 1 (January 1968): 22–43.

———. "The Nature and Diffusion of Technology." In A. J. Youngson, ed. *Economic Development in the Long Run*. London, 1972. Pp. 36–61.

———. *Technological Change: The United States and Great Britain in the 19th century*. London, 1970.

Schuyler, Eugene. *Selected Essays with a Memoir by Evelyn Schuyler Schaeffer*. New York, 1901.

Selwyn, J. N. "Breech-loaders, with Reference to Calibre, Supply and Cost of Ammunition." *Journal of the Royal United Services Institution* 11, no. 43 (1867): 15–26.

Sestroretskii instrumental'nyi zavod: Ocherki, dokumenty, vospominaniia, 1721–1967 (The Sestroretsk tool plant: essays, documents, and reminiscences, 1721–1967). *Istoriia fabrik i zavodov* series, no. 386. Leningrad, 1968.

Sharpe, Philip B. *The Rifle in America*. New York, 1938.

Shlakman, Vera. *Economic History of a Factory Town: A Study of Chicopee, Massachusetts*. Smith College Studies in History, 20. Northampton, Mass., 1935.

Showalter, Dennis. *Railroads and Rifles*. Hamden, Conn., 1975.

Shumilov, I. P. *Istoriia goroda Tuly i Imperatorskogo oruzheinogo zavoda* (A history of Tula and the Imperial armory). Tula, 1889.

Singer, Charles Joseph, ed. *A History of Technology*. 7 vols. Oxford, 1958.

Sirota, Y. I. "Perevooruzhenie russkoi armii vo II poloviny XIX v." (The rearmament of the Russian army during the second half of the 19th century). Candidate's diss. Leningrad, 1950.

Siscoe, F. G. "Eugene Schuyler, General Kaufman and Central Asia." *Slavic Review* 27, 1 (March 1968): 119–30.

Skalon, D., comp. *Stoletie Voennogo ministerstva, 1802–1902* (A century of the ministry of war, 1802–1902). 34 vols. St. Petersburg, 1902–1906.

Smiles, Samuel, ed. *James Nasmyth: Engineer.* New York, 1883.

Smith, A. L. "The Russian Army in the Balkans, 1877–1878." In J. G. Purves and D. A. West, eds. *War and Society in the 19th-century Russian Empire.* Toronto, 1972. Pp. 151–62.

"Smith and Wesson's Revolver Factory." *Scientific American* 42, 4, 24 January 1880.

Smith, Merritt Roe. *The Harper's Ferry Armory and the New Technology: The Challenge of Change.* Ithaca, N.Y., 1977.

Smith, Merritt Roe, ed. *Military Enterprise and Technological Change: Perspectives on the American Experience.* Cambridge, Mass., 1985.

Smith, W. H. B. *Book of Pistols and Revolvers.* 6th ed. Harrisburg, Pa., 1965.

———. *Small Arms of the World.* 10th ed. Harrisburg, Pa., 1973.

Solov'ev, A. *V pamiat' stoletiia iubileia osnovaniia Izhevskogo oruzheinogo zavoda* (In honor of the centenary of the Izhevsk small arms factory). Izhevsk, 1907.

Sovetskaia voennaia entsiklopediia (The Soviet military encyclopedia). 16 vols. Moscow, 1961–1971.

Stoletnaia godovshchina pribytiia russkikh eskadr v Ameriku (The centenary of the landing of the Russian squadron in America). Washington, 1963.

Strokov, A. A. *Istoriia voennogo iskusstva* (A history of the military arts). 2 vols. Moscow, 1965.

Strukov, D. P. *Arkhiv russkoi artillerii* (The archives of Russian artillery). Vol. 1. St. Petersburg, 1889.

———. *Putevoditel' po Artilleriiskomu istoricheskomu muzeiu* (Guidebook to the Artillery-Historical Museum). St. Petersburg, 1912.

Strumilin, S. G. *Istoriia chernoi metallurgii v SSSR* (A history of ferrous metallurgy in the USSR). Moscow, 1954.

———. *Promyshlennyi perevorot v Rossii* (The industrial revolution in Russia). Moscow, 1944.

Subbotin, Iu. F. "Iz istorii voennoi promyshlennosti Rossii kontsa XIX-nachala XX v." (Military industry in Russia at the end of the 19th and beginning of the 20th centuries). *VLU* 3 (1973): 45–51.

———. "K istorii vzaimootnoshenii mezhdu chastnymi i kazennymi predpriiatiiami v voennoi promyshlennosti Rossii kontsa XIX-nachale XX vv." (A history of the relationship between private and government establishments in Russian military industry at the end of the 19th and beginning of the 20th centuries). *VLU* 14 (1974): 35–42.

Sukinskii, A. "O novom ruzheinom porokhe dlia malokalibernoi vintovki" (New powder for the small-caliber rifle). *O.S.* 1 (1878): 31–83.

Sutton, A. C. *Western Technology and Soviet Economic Development.* 3 vols. Stanford, 1968–1973.

Tarassuk, Leonid. *The 'Russian' Colts: From Colonel Samuel Colt to the Russian Imperial Court.* North Hollywood, Calif., 1979.

———. *Russian Pistols in the 17th century.* York, Pa., 1968.

Tarassuk, Leonid, and Claude Blair, eds. *The Complete Encyclopedia of Arms and Weapons*. New York, 1979.

Tarassuk, Leonid, and R. L. Wilson. "The Russian Colts." *The Arms Gazette*, 3 installments (August-October 1976).

Tarsaidze, Alexander. "Berdanka." *Russian Review* 4, 1 (January 1950): 30–36.

———. *Czars and Presidents: The Forgotten Friendship*. New York, 1958.

Taylerson, A. W. F. *The Revolver, 1865–1888*. New York, 1956.

Tegoborski, Ludwik. *Commentaries on the Productive Forces of Russia*. 2 vols. London, 1855.

Temin, Peter. "Labor Scarcity and the Problem of American Industrial Efficiency in the 1850s." *Journal of Economic History* (September 1966): 277–98.

Thomas, Benjamin Platt. *Russo-American Relations, 1815–1867*. Baltimore, 1930.

Train, George Francis. *An American Merchant in Europe*. New York, 1857.

———. *My Life in Many States and in Foreign Lands*. New York, 1902.

Trebilcock, C. "Spin-Off in British Economic History: Armaments and Industry, 1760–1914." *Economic History Review* 22 (December 1969): 474–90.

———. "War and the Failure of Industrial Modernization, 1899–1914." In J. M. Winter, ed. *War and Economic Development: Essays in Memory of David Joslin*. Cambridge, 1975.

Trutnev, N. F. "Tul'skaia oruzheinaia sloboda i kazennyi zavod v pervoi chetverti XVIII v." (The Tula armorers' settlement and the state factory during the first quarter of the 18th century). In V. N. Ashurkov, ed. *Iz istorii Tuly i Tul'skogo kraia* (The history of Tula and the Tula region). Tula, 1983. Pp. 113–30.

Tsybul'skii, V. A. "Sestroretskii oruzheinyi zavod i perevooruzhenie russkoi armii v kontse XIX v." (The Sestroretsk small arms factory and the rearmament of the Russian army at the end of the 19th century). In V. V. Mavrodin, ed. *Rabochie oruzheinoi promyshlennosti v Rossii i russkie oruzheiniki v XIX-nachale XX v.* Leningrad, 1976. Pp. 60–69.

Tweedale, Geoffrey. "Sheffield Steel and America: Aspects of the Atlantic Migration of Special Steelmaking Technology, 1850–1930." *Business History* 25, 3 (November 1983): 225–39.

Urusov, S. S. *Ocherki vostochnoi voiny, 1854–55* (Studies on the Eastern War, 1854–55). Moscow, 1866.

U.S. Congress. House Committee on International Relations, Subcommittee on International Security and Scientific Affairs. *Technology transfer and scientific cooperation between the United States and the Soviet Union: A Review*. 95th Cong., 1st sess. Washington, D.C., 1977.

Uselding, Paul. "An Early Chapter in the Evolution of American Industrial Management." In Louis P. Cain and Paul Uselding, eds. *Business Enterprise and Economic Change*. Kent, Ohio, 1973. Pp. 51–84.

———. "Elisha K. Root and the American System." *T.C.* 15, 4 (October 1974): 543–68.

------. "Henry Burden and the Question of Anglo-American Technology Transfer in the 19th century." *Journal of Economic History* 30 (1970): 312–37.

------. "Studies of Technology in Economic History." In Robert E. Gallman, ed. *Recent Developments in the Study of Business Economic History: Essays in Memory of Herman E. Krooss.* Greenwich, Conn., 1977.

Van Slyck. *Representatives of New England Manufacturers.* 2 vols. Boston, 1879.

Vasil'chikov, V. I. "Zapiski o tom, pochemu russkoe oruzhie postoianno terpelo neudachu i na Dunae i v Krymu v 1853–55 gg." (Why Russian firearms repeatedly failed on the Danube and in the Crimea from 1853 to 1855). *Russkii arkhiv* 6 (1891): 167–256.

Vasiliev, A. P. "K istorii Izhevskogo chastnogo oruzheinogo proizvodstva, 1867–1917" (The history of Izhevsk private arms production, 1867–1917). In *Voprosy istorii razvitiia promyshlennosti Udmurtrii, 1861–1985: Sbornik statei* (Problems in the development of industry in Udmurtriia, 1861–1985: a collection of articles). Ustinov, 1986. Pp. 87–99.

Vel'iaminov-Zernov, A. "Teoreticheskii kurs o ruchnom ognestrel'nom oruzhii, prepodavaemyi pri uchebnom pekhotnom batalione" (Theoretical course on firearms, taught at the infantry training battalion). *O.S.*, 4 installments (1864).

Verkovskii, A. *Ocherk po istorii voennogo iskusstva v Rossii XVII-XIX vv.* (A study of the history of military art in Russia in the 18th and 19th centuries). Moscow, 1921.

Vershigora, P. P. *Voennoe tvorchestvo narodnykh mass* (A people's military creativity). Moscow, 1961.

Veshniakov, V. "Russkaia promyshlennost' i ee nuzhdy" (The needs of Russian industry). *Vestnik Evropy* 10–12 (1870).

Vincent, C. E. H. "Russian Army." *Journal of the Royal United Services Institution* 16, 67 (1872): 285–308.

Virginskii, V. S. *Tvortsy novoi tekhniki v krepostnoi Rossii* (Inventors in serf Russia). 2d ed. Moscow, 1962.

Viskovatov, A. V. *Istoricheskoe opisanie odezhdy i vooruzheniia, 1841–62* (Historical description of uniforms and weapons, 1841–62). 30 vols. St. Petersburg, 1841–1862.

Voennyi entsiklopedicheskii slovar' (A military encyclopedic dictionary). 2d ed. Moscow, 1986.

Voennaia entsiklopediia (Military encyclopedia). vols. 1–18. St. Petersburg, 1911–1915.

Von Wahlde, Peter. "Military Thought in Imperial Russia." Ph.D. diss. Indiana University, 1966.

------. "Russian Military Reform, 1862–1874." *Military Review* 39, 10 (January 1960): 60–69.

Voprosy voennoi istorii Rossii XVIII v. i pervoi poloviny XIX v. (Problems in Russian military history of the 18th and first half of the 19th centuries). Moscow, 1969.

Vrubel', V. "Vizit druzhby russkikh korablei k beregam Ameriki" (The friendly visit of the Russian ships to American shores). *Voenno-Istoricheskii Zhurnal* 4 (1977): 101–5.

Wahl, Paul, and Donald R. Toppel. *The Gatling Gun.* New York, 1965.

Weigley, Russell F., ed. *New Dimensions in Military History.* San Rafael, Calif., 1975.

Weller, Jac. *Weapons and Tactics.* New York, 1966.

Wellesley, Col. F. A. *With the Russians in Peace and War.* London, 1905.

White, Andrew D. *Autobiography.* 2 vols. New York, 1905.

Wilferd, Col. E. C. "On the progress of fire-arms for military purposes to their present state." *Journal of the Society of Arts* (11 May 1866): 439–45.

Wilkins, Mira. *The Emergence of Multinational Enterprise: American Business Abroad from the Colonial Era to 1914.* Cambridge, Mass., 1970.

Williamson, Harold F. *Winchester, the Gun that won the West.* Washington, 1952.

Wilson, R. L. *Colt, An American Legend: The Official History of Colt Firearms from 1836 to the Present.* New York, 1985.

———. *Samuel Colt Presents.* New York, 1961.

Winchester Repeating Arms Company. New Haven, 1869.

Winter, J. M., ed. *War and Economic Development: Essays in Memory of David Joslin.* Cambridge, 1975.

Wintringham, T. *The Story of Weapons and Tactics from Troy to Stalingrad.* London, 1943.

Woodbury, Robert S. *History of the Milling Machine.* Cambridge, Mass., 1960.

———. *Studies in the History of Machine Tools.* Cambridge, Mass., 1972.

Zagorskii, F. N. *A History of Metal cutting machines to the middle of the 19th century.* Trans. Edwin Battison. New Delhi, 1982.

———. *L. F. Sabakin, A Russian Mechanic of the 18th century: His Life and Work.* Jerusalem, 1966.

Zaionchkovskii, A. M. "Vzgliad N. N. Murav'eva na sostoianie nashei armii i sobstvennoruchnye zamechaniia Imp. Nikolaia Pavlovicha" (N. N. Muravev's view of the condition of our army and the hand-written comments of Emperor Nicholas I) *V.S.* 2 (1903): 255–68.

Zaionchkovskii, P. A. "Perevooruzhenie russkoi armii v 60–70kh gg. XIX v." (The rearmament of the Russian army, 1860–1870). *Istoricheskie zapiski* 36 (1951): 64–100.

———. "Voennye reformy D. A. Miliutina" (Miliutin's military reforms). *Voprosy istorii* 2 (1945): 3–27.

———. *Voennye reformy, 1860–1870* (Military reforms, 1860–1870). Moscow, 1952.

Zaleski, E., and H. Wienert. *Technology Transfer between East and West.* Paris, 1980.

Zelentsov, V. D. "Otmena krepostnogo truda na Izhevskom zavode v 1866 g." (The abolition of serf labor at the Izhevsk factory in 1866). *Uchenye zapiski Gor'kovskogo pedagogicheskogo instituta* 97 (1934): 105–9.

————. "Rabochie Izhevskogo zavoda v dokapitalisticheskuiu epokhu" (Workers at the Izhevsk factory in the pre-capitalist era). *Istoriia proletariata* 4 (1934): 73–92.

Zhilin, M. A. *Problemy voennoi istorii* (Problems of military history). Moscow, 1975.

Index